EVALUATING CHEMICAL
AND OTHER AGENT
EXPOSURES
FOR
REPRODUCTIVE
AND
DEVELOPMENTAL
TOXICITY

Subcommittee on Reproductive and Developmental Toxicology

Committee on Toxicology

Board on Environmental Studies and Toxicology

Commission on Life Sciences

National Research Council

NATIONAL ACADEMY PRESS
Washington, D.C.

NATIONAL ACADEMY PRESS   2101 Constitution Ave., N.W.   Washington, D.C. 20418

NOTICE: The project that is the subject of this report was approved by the Governing Board of the National Research Council, whose members are drawn from the councils of the National Academy of Sciences, the National Academy of Engineering, and the Institute of Medicine. The members of the committee responsible for the report were chosen for their special competences and with regard for appropriate balance.

This project was supported by Contract Nos. DAMD 17-89-C9086 and DAMD 17-99-C9049 between the National Academy of Sciences and the U.S. Department of Defense. Any opinions, findings, conclusions, or recommendations expressed in this publication are those of the author(s) and do not necessarily reflect the view of the organizations or agencies that provided support for this project.

International Standard Book Number 0-309-07316-2

Additional copies of this report are available from:

National Academy Press
2101 Constitution Ave., NW
Box 285
Washington, DC 20055

800-624-6242
202-334-3313 (in the Washington metropolitan area)
http://www.nap.edu

THE NATIONAL ACADEMIES

National Academy of Sciences
National Academy of Engineering
Institute of Medicine
National Research Council

The **National Academy of Sciences** is a private, nonprofit, self-perpetuating society of distinguished scholars engaged in scientific and engineering research, dedicated to the furtherance of science and technology and to their use for the general welfare. Upon the authority of the charter granted to it by the Congress in 1863, the Academy has a mandate that requires it to advise the federal government on scientific and technical matters. Dr. Bruce M. Alberts is president of the National Academy of Sciences.

The **National Academy of Engineering** was established in 1964, under the charter of the National Academy of Sciences, as a parallel organization of outstanding engineers. It is autonomous in its administration and in the selection of its members, sharing with the National Academy of Sciences the responsibility for advising the federal government. The National Academy of Engineering also sponsors engineering programs aimed at meeting national needs, encourages education and research, and recognizes the superior achievements of engineers. Dr. William A. Wulf is president of the National Academy of Engineering.

The **Institute of Medicine** was established in 1970 by the National Academy of Sciences to secure the services of eminent members of appropriate professions in the examination of policy matters pertaining to the health of the public. The Institute acts under the responsibility given to the National Academy of Sciences by its congressional charter to be an adviser to the federal government and, upon its own initiative, to identify issues of medical care, research, and education. Dr. Kenneth I. Shine is president of the Institute of Medicine.

The **National Research Council** was organized by the National Academy of Sciences in 1916 to associate the broad community of science and technology with the Academy's purposes of furthering knowledge and advising the federal government. Functioning in accordance with general policies determined by the Academy, the Council has become the principal operating agency of both the National Academy of Sciences and the National Academy of Engineering in providing services to the government, the public, and the scientific and engineering communities. The Council is administered jointly by both Academies and the Institute of Medicine. Dr. Bruce M. Alberts and Dr. William A. Wulf are chairman and vice chairman, respectively, of the National Research Council.

SUBCOMMITTEE ON REPRODUCTIVE AND DEVELOPMENTAL TOXICOLOGY

CAROLE A. KIMMEL *(Chair)*, U.S. Environmental Protection Agency, Washington, D.C.

GERMAINE M. BUCK, University at Buffalo, State of New York

MAUREEN H. FEUSTON, Sanofi Pharmaceuticals, Inc., Malvern, Pa.

PAUL M.D. FOSTER, Chemical Industry Institute of Toxicology, Research Triangle Park, N.C.

J. M. FRIEDMAN, University of British Columbia, Vancouver

JOSEPH F. HOLSON, WIL Research Laboratories, Inc., Ashland, Ohio

CLAUDE L. HUGHES, JR., Cedars-Sinai Medical Center, Los Angeles, Calif.

JOHN A. MOORE, Institute for Evaluating Health Risks, Washington, D.C., and Sciences International, Alexandria, Va.

BERNARD A. SCHWETZ, National Center for Toxicological Research, Rockville, Md.

ANTHONY R. SCIALLI, Georgetown University Medical Center, Washington, D.C.

WILLIAM J. SCOTT, JR., University of Cincinnati College of Medicine, Cincinnati, Ohio

CHARLES V. VORHEES, University of Cincinnati College of Medicine, Cincinnati, Ohio

BARRY R. ZIRKIN, The Johns Hopkins University School of Hygiene and Public Health, Baltimore, Md.

Staff

KULBIR S. BAKSHI, Program Director for the Committee on Toxicology

ABIGAIL STACK, Project Director

KATE KELLY, Editor

MIRSADA KARALIC-LONCAREVIC, Information Specialist

LEAH PROBST, Senior Project Assistant

EMILY SMAIL, Project Assistant

Sponsor

U.S. DEPARTMENT OF DEFENSE

COMMITTEE ON TOXICOLOGY

Staff

KULBIR S. BAKSHI, Program Director
SUSAN N.J. MARTEL, Program Officer
ABIGAIL E. STACK, Program Officer
RUTH E. CROSSGROVE, Publications Manager
KATHRINE J. IVERSON, Manager, Toxicology Information Center
AIDA C. NEEL, Senior Project Assistant
EMILY L. SMAIL, Project Assistant

Hazardous Materials on the Public Lands (1992)
Science and the National Parks (1992)
Animals as Sentinels of Environmental Health Hazards (1991)
Assessment of the U.S. Outer Continental Shelf Environmental
    Studies Program, Volumes I-IV (1991-1993)
Human Exposure Assessment for Airborne Pollutants (1991)
Monitoring Human Tissues for Toxic Substances (1991)
Rethinking the Ozone Problem in Urban and Regional Air Pollution
    (1991)
Decline of the Sea Turtles (1990)

Copies of these reports may be ordered from
the National Academy Press
(800) 624-6242
(202) 334-3313
www.nap.edu

Acute Exposure Guideline Levels for Selected Airborne
    Contaminants, Volume 1 (2000)
Review of the US Navy's Human Health Risk Assessment of the
    Naval Air Facility at Atsugi, Japan (2000)
Methods for Developing Spacecraft Water Exposure Guidelines
    (2000)
Review of the U.S. Navy Environmental Health Center's Health-
    Hazard Assessment Process (2000)
Review of the U.S. Navy's Exposure Standard for Manufactured
    Vitreous Fibers (2000)
Re-Evaluation of Drinking-Water Guidelines for Diisopropyl
    Methylphosphonate (2000)
Submarine Exposure Guidance Levels for Selected
    Hydrofluorocarbons: HFC-236fa, HFC-23, and HFC-404a (2000)
Review of the U.S. Army's Health Risk Assessments for Oral
    Exposure to Six Chemical-Warfare Agents (1999)
Toxicity of Military Smokes and Obscurants, Volume 1(1997),
    Volume 2 (1999), Volume 3 (1999)
Assessment of Exposure-Response Functions for Rocket-Emission
    Toxicants (1998)
Toxicity of Alternatives to Chlorofluorocarbons: HFC-134a and
    HCFC-123 (1996)
Permissible Exposure Levels for Selected Military Fuel Vapors (1996)
Spacecraft Maximum Allowable Concentrations for Selected
    Airborne Contaminants, Volume 1 (1994), Volume 2 (1996),
    Volume 3 (1996), Volume 4 (2000)

Preface

The United States Navy has been concerned for some time with protecting its military and civilian personnel from reproductive and developmental hazards in the workplace. As part of its efforts to reduce or eliminate exposure of Naval personnel and their families to reproductive and developmental toxicants, the Navy requested that the National Research Council (NRC) recommend an approach that can be used to evaluate chemicals and physical agents for their potential to cause reproductive and developmental toxicity. The NRC assigned this project to the Committee on Toxicology, which convened the Subcommittee on Reproductive and Developmental Toxicology, to prepare this report. In this report, the subcommittee recommends an approach for evaluating agents for potential reproductive and developmental toxicity and demonstrates how that approach can be used by the Navy.

Several individuals assisted the subcommittee by providing information on Naval operations, particularly on the Navy's health hazard evaluation program. We thank Captain David Macys (Office of Naval Research), Captain Lawrence Betts (Navy Environmental Health Center), Commander Victoria Cassano (Navy Environmental Health Center), Captain David Sack (Navy Environmental Health Center), Captain Kenneth Still (Navy Health Research Center's Toxicology Detachment), and James Crawl (Navy Environmental Health Center) for their interest and support of this project. We also gratefully acknowledge the following persons who provided valuable background information to the subcommittee: Stacy Arnesen (National Library of Medicine), George Daston (Procter and Gamble Company), James Donald (California Environmental Protection Agency), Elaine Faustman (University of Washington), Michael Shelby (National Institute of Environmental Health Sciences), and John Weiner (University at Buffalo, State of New York). The subcommittee thanks R. Woodrow Setzer, Jr. (U.S. Environmental Protection Agency) for providing guidance on statistical methods discussed in this report.

This report has been reviewed in draft form by individuals chosen for their diverse perspectives and technical expertise, in accordance with procedures approved by the NRC's Report Review Committee. The purpose of this independent review is to provide candid and critical comments that will assist the institution in making its published report as sound as possible and to ensure that the report meets institutional standards for objectivity, evidence, and responsiveness to the study charge. The review comments and draft manuscript remain confidential to protect the integrity of the deliberative process. We wish to thank the following individuals for their review of this report: James Chen (National Center for Toxicological Research), George Daston (Procter and Gamble Company), Jerry Heindel (National Institute of Environmental Health Sciences), Grace Lemasters (University of Cincinnati), and John Young (National Center for Toxicological Research).

Although the reviewers listed above have provided many constructive comments and suggestions, they were not asked to endorse the conclusions or recommendations, nor did they see the final draft of the report before its release. The review of this report was overseen by Donald Mattison (March of Dimes Birth Defects Foundation), appointed by the Commission on Life Sciences, who was responsible for making certain that an independent examination of this report was carried out in accordance with institutional procedures and that all review comments were carefully considered. Responsibility for the final content of this report rests entirely with the authoring committee and the institution.

We are also grateful for the assistance of NRC staff in the preparation of this report. The subcommittee acknowledges Kulbir Bakshi, program director of the Committee on Toxicology and, in particular, Abigail Stack, project director for this report, without whose leadership and assistance this project could not have been completed. Other staff members contributing to this report were James Reisa, director of the Board on Environmental Studies and Toxicology (BEST); Carol Maczka, formerly BEST's senior program director for toxicology and risk assessment; Ruth Crossgrove, publications manager; Leah Probst, senior project assistant; and Emily Smail, project assistant.

Finally, we thank all the members of the subcommittee for their expertise and dedicated effort throughout the study.

Carole A. Kimmel, Ph.D.
Chair, Subcommittee on Reproductive
and Developmental Toxicology

Bailus Walker Jr., Ph.D., M.P.H.
Chair, Committee on Toxicology

Contents

Abbreviations

ACGIH	American Conference of Governmental Industrial Hygienists
ADI	acceptable daily intake
AIHA	American Industrial Hygiene Association
ATSDR	Agency for Toxic Substances and Disease Registry
AUC	area under the curve
BMD	benchmark dose
CAS	Chemical Abstract Service
CDC	Centers for Disease Control and Prevention
CFC	chlorofluorocarbon
Cmax	peak threshold concentration
DART	developmental and reproductive toxicology
EC	European Commission
ECETOC	European Centre for Ecotoxicology and Toxicology of Chemicals
EPA	U.S. Environmental Protection Agency
ETICBACK	Environmental Teratology Information Center Backfile
F_1	first filial generation
FDA	U.S. Food and Drug Administration
HEC	human equivalent concentration
HFC 134a	1,1,1,2-tetrafluoroethane
HFC	hydrofluorocarbon

HSDB	Hazardous Substance Data Base
IARC	International Agency for Research on Cancer
ILO	International Labor Organization
IPCS	International Programme on Chemical Safety
IRIS	Integrated Risk Information System (Administered by the U.S. Environmental Protection Agency)
JP-8	jet propellant-8
LC_{50}	lethal concentration for 50% of the test animals
LD_{50}	lethal dose for 50% of the test animals
LHRH	luteinizing hormone releasing hormone
LOAEL	lowest-observed-adverse-effect level
MDI	metered dose inhaler
MeSH	medical subject headings
MOE	margin of exposure
MTD	maximum tolerated dose
NCEA	National Center for Environmental Assessment
NIEHS	National Institute of Environmental Health Sciences
NIH	National Institutes of Health
NIOSH	National Institute for Occupational Safety and Health
NOAEL	No-observed-adverse-effect level
NRC	National Research Council
NTIS	National Technical Information Service
NTP	National Toxicology Program
OECD	Organization for Economic Cooperation and Development
OR	odds ratio
ORD	Office of Research and Development
P generation	parental animals
PEL	permissible exposure limit
RACB	reproductive assessment of continuous breeding
RfC	reference concentration
RfD	reference dose
RR	relative risk
RTECS	Registry of Toxic Effects of Chemical Substances
SIDS	screening information data set
SOP	standard operating procedure
STEL	Short-Term Exposure Limit

TLV	Threshold Limit Value
TOXNET	Toxicology Data Network
UEL	unlikely effect level
UF	uncertainty factor
UNEP	United Nations Environmental Program
WHO	World Health Organization

EVALUATING CHEMICAL
AND OTHER AGENT
EXPOSURES
FOR
REPRODUCTIVE
AND
DEVELOPMENTAL
TOXICITY

Summary

Reproductive disorders and developmental defects (including birth defects) are significant public health problems, with enormous personal and economic costs. Reproductive disorders may include altered menstrual and ovarian cycles, increased time-to-pregnancy, decreased sperm count, reduced libido, and infertility. Developmental defects may be manifested as prenatal and postnatal death, structural abnormalities (e.g., neural tube and heart defects), altered growth (e.g., low birth weight), and functional deficiencies (e.g., mental retardation). The known causes of reproductive and developmental disorders include genetic mutations; maternal metabolic imbalances; infection; and occupational, therapeutic, and environmental exposure to harmful chemical and physical agents.

Concern regarding reproductive and developmental hazards in the workplace, including military facilities, has increased significantly in recent years. In 1997, Congress passed a law, as part of the National Defense Authorization Act, concerning health care coverage for children with medical conditions caused by parental exposure to hazardous materials while serving as members of the Armed Services (Public Law 104-201, Section 704). The law states, in part, that a plan would be developed for ensuring the provision of medical care to any natural child of a member of the Armed Forces who has a congenital defect or catastrophic illness, proven to a reasonable degree of scientific certainty on the basis of scientific research to have resulted from exposure of the member to a chemical warfare agent or other hazardous material to which the member was exposed during active military service. The Department of Defense is required to develop a plan for compliance.

As a part of its efforts to protect military and civilian personnel from reproductive and developmental hazards in the workplace, the Navy requested that the National Research Council (NRC) recommend an approach that can be used to evaluate sources of potential reproductive and developmental toxicity. The NRC assigned this project to the Committee on Toxicology, which convened the Subcommittee on Reproductive and Developmental Toxicology. The subcommittee was assigned the following tasks:

- Develop a process for assessing the reproductive and developmental toxicity potential from exposures to chemicals and physical agents.
- Develop a strategy for dealing with the potential reproductive and developmental toxicity of exposures to chemicals and physical agents for which little or no information is available.
- Conduct pilot evaluations on two chemicals using the process developed by the subcommittee.
- Identify reliable sources for assessment of reproductive and developmental toxicity.
- Identify areas of needed research.

In this report, the subcommittee recommends an approach to assess potential reproductive and developmental toxicity from exposures to substances encountered in workplaces operated by the United States Navy.

CONCLUSIONS AND RECOMMENDATIONS

The subcommittee's major conclusions and recommendations, organized in response to each of its tasks, are presented below.

Evaluative Process

The subcommittee's recommended approach for evaluating exposures to chemicals and physical agents for reproductive and developmental toxicity is based on a process published in 1995 by Moore and colleagues.[1] As it is described and expanded on in the subcommittee's report, the process undertakes a systematic review of data on reproductive and developmental toxicity in humans and experimental animals, on general toxicity, and on the conditions of use that result in human exposure. The toxicity and exposure data are integrated, and the result is an estimate of an exposure that is unlikely to cause reproductive or developmental toxicity.

The subcommittee recommends against an attempt to classify agents as "toxic" or "nontoxic." Instead, the potential toxicity of a substance should be considered in the context of exposure (e.g., amount, route, timing, and duration of exposure). The subcommittee recognizes that the Navy might want to use a screening process in which decisions are made in a dichotomous manner (to use or not to use a particular agent). Such decisions can be made by considering the exposure scenario that is anticipated in the workplace. An exposure level of an agent that is unlikely to be associated with reproductive and developmental toxicity can be estimated, and if the anticipated workplace exposure is sufficiently lower than that estimate, the Navy can regard the exposure as acceptable. If the anticipated human exposure is higher than the estimate, then the use of the agent under consideration can be regarded as unacceptable and exposure control measures can be implemented or alternative agents can be evaluated.

The subcommittee recommends that the evaluative process be implemented by a team of scientists with training and experience in assessing agents for their potential to cause reproductive and developmental toxicity. Such evaluation requires expertise in the intricacies and relationships of the integrated processes of reproduction and development. Considerable scientific judgment is needed to interpret data and make informed decisions about the adequacy of available data sets for estimating the potential reproductive and developmental toxicity of specific substances under specific conditions of exposure.

---

[1]Moore, J.A., G.P. Daston, E. Faustman, M.S. Golub, W.L. Hart, C. Hughes Jr., C.A. Kimmel, J.C. Lamb IV, B.A. Schwetz, and A.R. Scialli. 1995. An evaluative process for assessing human reproductive and developmental toxicity of agents. Reprod. Toxicol. 9(1):61-95.

Assessing the Available Data

When conducting an evaluation of an agent for potential reproductive and developmental toxicity, the Navy should assess several types of data: human exposure data, general toxicity data in humans and experimental animals, and reproductive and developmental toxicity data in humans and experimental animals. Complete assessments should consider potential adverse effects on the male and female reproductive systems and on the embryo, fetus, and child.

Human exposure data are evaluated to identify populations that might be exposed, to identify potential pathways of exposure, and to estimate the range of exposure so that quantitative estimates of exposure can be made that are associated with each pattern of use. Exposure conditions that are unique for reproductive and developmental toxicity should be considered because the embryo, fetus, neonate, juvenile, young adult, and older adult differ in susceptibility. Human exposure data are important for accurate evaluation of the risk potential of an agent, but data of sufficient quality and quantity are often unavailable.

Chemical data (e.g., physical and chemical properties, structure-activity relationships, and environmental fate and transport), basic toxicity data, and pharmacokinetic data (information on absorption, distribution (including placental and lactational transfer), metabolism, and excretion) should be reviewed. These data are particularly important because reproductive and developmental effects are interpreted in the context of general toxicity data in humans or experimental animals. Pharmacokinetic data for both animals and humans can be helpful in extrapolating exposure levels from one species to another.

Reproductive and developmental toxicity data from animal experiments and human studies should be assessed based on defined criteria. One of the following judgments can be made: either the toxicity data are sufficient (or insufficient) to ascribe an adverse effect to a specific agent under specified conditions, or the data are sufficient (or insufficient) to conclude that there is no adverse effect. To be characterized as sufficient, the database must include information on the full range of potential adverse male and female reproductive effects and developmental effects, and the actual range of conditions of exposure must be known in sufficient detail to determine whether the dose, duration, route, timing, and other characteristics of exposure pose a substantial reproductive risk. A designation of sufficient (or insufficient) data is inadequate by itself to identify a substance as having the potential to cause (or not cause) reproductive or developmental toxicity; the reproductive and developmental toxicity information must be integrated with the exposure and general toxicity information before the evaluative process can be considered complete. The integration step is described below.

Integration of Toxicity and Exposure Information

The integration step of the evaluation is conducted in three stages. In the first stage, the evaluators examine the data for relevance to potential human toxicity. Then, if the data are determined to be relevant to human exposures, a quantitative assessment is conducted. Finally, the concluding step of the evaluative process is the integration of toxicity and exposure information to characterize the risk of potential reproductive and developmental toxicity.

This step involves combining information from the review of animal and human reproductive and developmental toxicity data with information from the review of general toxicity, pharmacokinetic, and exposure data. A weight-of-evidence approach is used to formulate judgments about potential hazards to humans. Three separate judgments should be developed: one each to address developmental toxicity, female reproductive toxicity, and male reproductive toxicity.

Once an assessment has determined that the data indicate human risk potential for reproductive and developmental toxicity, the next step is to perform a quantitative evaluation. Dose-response data from human and experimental animal reproductive and developmental toxicity studies are reviewed to identify a no-observed-adverse-effect level (NOAEL) or a lowest-observed-adverse-effect level (LOAEL), and/or to derive a benchmark dose (BMD). Duration adjustments of the NOAEL, LOAEL, or BMD are often made, particularly for inhalation exposures when extrapolating to different exposure scenarios. Such adjustments have not been routinely applied to developmental toxicity data. The subcommittee recommends that duration adjustments be considered for both reproductive and developmental toxicity assessments. Uncertainty factors (UF) are then applied to the NOAEL, LOAEL, or BMD to account for various uncertainties in the data. Uncertainty factors for reproductive and developmental toxicity commonly applied to the NOAEL or BMD include a 10-fold factor for interspecies variability and a 10-fold factor for intraspecies variability. If only a LOAEL is available, an additional factor of up to 10-fold would be applied. The magnitude of the UFs can be adjusted, depending on the type and quantity of data, including pharmacokinetic and pharmacodynamic data, available and other modifying factors can be used to account for other uncertainties (e.g., insufficiencies in the database).

To calculate an unlikely effect level (UEL) for reproductive and developmental toxicity, the NOAEL, LOAEL, or BMD is divided by the composite UF. UELs can be calculated for different exposure durations or adjusted to account for length of exposure. A UEL can be compared with a human exposure estimate to determine whether the exposure is sufficient to cause concern. If the UEL is higher than the human exposure estimate, there will be little or no cause for concern. If the UEL is lower than the human exposure estimate, then there is a possibility that adverse effects may occur. A margin of exposure (MOE; the ratio of the NOAEL or BMD to the anticipated human exposure) also can be calculated. In that case, the higher the ratio, the greater the numerical distance between the human exposure estimate and the highest dose that is without adverse effect in the species tested. The choice of an MOE for regulatory action should be based on the level of confidence in the underlying data and on judgment about other factors that might influence human risk, similar to the judgment made in the selection of appropriate UFs. It would be inappropriate to use a particular MOE as a default action level.

Each evaluation should conclude with a summary of the risk posed by a substance. The summary can consist of background information on the chemical and toxicological parameters of the agent; human exposure information; a summary of the male and female reproductive toxicity data and the developmental toxicity data; a list of the quantitative values derived from the data; a description of the default assumptions and UFs used in the process; the data needs to reduce uncertainty; and a reference section.

Insufficient Data Sets

In practice, sufficient data are rarely available to inform a judgment about the potential reproductive or developmental toxicity of an exposure to an agent. In such cases, steps can be taken to minimize the risk of an adverse effect. Obviously, the only way to completely eliminate the risk of an adverse effect is to eliminate exposure to the agent, but that is often not feasible. If use of a given agent is unavoidable, the risk can be minimized by assuming that susceptibility to reproductive and developmental toxicity may be greater than susceptibility to any other known toxicity of the agent and applying additional UFs to reflect the lack of data. The risk can also be minimized by substituting an agent that is known not to be associated with substantial reproductive or developmental toxicity or by limiting the potentially absorbed dose by the use of respirators, gloves, and protective clothing.

Application of the Evaluative Process

To demonstrate how the subcommittee's recommended evaluative process can be applied to specific agents, the subcommittee evaluated two compounds of interest to the Navy: jet propulsion fuel 8 (JP-8) and hydrofluorocarbon (HFC) 134a. These assessments demonstrate that the subcommittee's recommended process can be used to evaluate compounds for which varying amounts of data are available. For example, several reproductive and developmental toxicity studies have been conducted for HFC 134a; however, just one developmental toxicity study has been conducted for JP-8. The subcommittee calculated a UEL based on at least one endpoint for each compound, accounting for uncertainties due to deficiencies in the database. Regardless of the quantity of data available, the subcommittee found that considerable scientific judgment was needed to conduct the evaluations.

Although these assessments are examples of how to use the evaluative process, they are not complete — largely because the subcommittee was unable to assess thoroughly the conditions of use that result in human exposure. For example, JP-8 is a complex chemical mixture and although the subcommittee did evaluate toxicity data for the fuel, it did not evaluate the constituents separately. The Navy should evaluate data on the individual components of JP-8 because the composition of the fuel mixture can vary from batch to batch. The subcommittee recommends that the Navy examine how various environmental conditions (e.g., temperature extremes and humidity) might affect exposure patterns to HFC 134a and JP-8.

Sources of Information

To assist the Navy in gathering information on chemicals and physical agents, the subcommittee reviewed many of the information sources available on reproductive and developmental toxicology. For most exposures, no single source includes all the information needed for a comprehensive evaluation of reproductive and developmental toxicology, and it is usually necessary to review several different sources of information.

The subcommittee reviewed sources of information that are specifically designed to assess reproductive and developmental toxicity and sources of information that are not as specific, but often contain some information. The subcommittee's summaries briefly describe the type of information provided by each source, the quality-control procedures (e.g., peer review), and how useful that source is in identifying exposures that pose a risk of reproductive and developmental toxicity in humans.

Sources specific to reproductive and developmental toxicity include detailed evaluations on specific agents (e.g., the California Environmental Protection Agency Hazard Identification Documents on Reproductive and Developmental Toxicity), informational summaries on specific agents that are not as comprehensive as the detailed evaluations (e.g., electronic databases such as REPROTEXT and reference books such as J.L. Schardein's *Chemically Induced Birth Defects*), bibliographic sources (e.g., the Developmental and Reproductive Toxicology Database (DART) maintained by the National Library of Medicine), and primary data (e.g., the National Toxicology Program reproductive and developmental toxicology study reports). The subcommittee believes that DART, which covers the literature on teratology and some aspects of reproductive and developmental toxicology, is, in particular, an essential resource to the Navy because it greatly simplifies the process for searching for literature in this area. Sources that are not specifically designed to evaluate reproductive and developmental toxicity, but contain some useful information on such toxicity (e.g., the Agency for Toxic Substances and Disease Registry Toxicological Profiles and the Environmental Protection Agency Integrated Risk Information System) are evaluated as well.

Research Recommendations

The subcommittee recommends that the Navy conduct research to obtain sufficient data sets on agents in use or considered for use. Specifically, the Navy should conduct the following tasks:

- Conduct experimental studies in animals to assess the potential for agents to cause reproductive and developmental effects on the male and female reproductive systems and on the embryo, fetus, and child.
- Fill data gaps that would reduce uncertainties in data sets and, thereby, eliminate or reduce the need for default uncertainty factors.
- Conduct toxicity studies on chemical mixtures.
- Design, implement, and conduct epidemiological studies that focus on various reproductive and developmental outcomes. Naval ships provide a unique environment in which to study a well-defined population—one in which many confounders that affect community or occupational studies (e.g., life-style factors thought to affect reproductive health such as alcohol consumption and cigarette smoking) can be documented.
- Conduct studies to estimate exposures and, in particular, consider exposure scenarios that are unique to the Navy's work environment.

1

Introduction

Disorders of human reproduction and development affect a substantial number of individuals of all ages and have long-standing implications for the public in terms of both quality of life issues and economic costs. Although reproductive disorders typically affect younger individuals, they also have implications for long-term health status. For example, men with impaired fecundity (the biological capacity for reproduction) could be at increased risk for testicular cancer (Moller and Skakkebaek 1999) and women with early onset of menopause could be at increased risk of osteoporosis (Bagur and Mautalen 1992; Ohta et al. 1996).

In assessing reproductive and developmental toxicity, a spectrum of endpoints in adult men and women, embryos, fetuses, infants, and children need to be considered. Reproductive outcomes are typically measures related to fertility (the ability to conceive) and fecundity. Adverse reproductive outcomes include abnormal male and female hormone profiles, altered menstrual and ovarian cycles, longer than normal time-to-pregnancy, abnormal semen characteristics, gynecological and urological disorders, spontaneous abortion, ectopic pregnancy (i.e., a pregnancy occurring elsewhere than in the uterus), and premature reproductive senescence. Major manifestations of abnormal development include pre- and postnatal deaths (e.g., spontaneous abortions, still births, and infant deaths), birth defects, altered growth, and functional deficiency (e.g., neurological, respiratory, and immune deficiencies) in the offspring (Wilson 1973).

Known causes of reproductive and developmental disorders include genetic defects; maternal metabolic imbalances; infection; and occupational, therapeutic, and environmental exposures to chemical and physical agents. This report primarily addresses reproductive and developmental defects that might be attributable to chemical and other agent occupational exposures.

The number of individuals affected by reproductive disorders is difficult to assess, and few population-based data are available for either men or women. Noticeably absent are data on fecundity and fertility impairments affecting men and only limited information on male-mediated developmental outcomes exists. Population-based data for impaired female fertility are available for select endpoints from the National Surveys of Family Growth (NSFG), which are conducted periodically and most recently in 1995. Data from the NSFG show that 6.2 million women (10.2%) between the ages of 15 and 44 in the United States had impaired fertility in 1995 (Stephen 1996). This number was estimated to increase to 6.3 million women in 2000 (Stephen and Chondra 1998). Other reproductive disorders in females that impact fecundity include endometriosis and polycystic ovarian syndrome (PCOS). The prevalence of endometriosis in women of reproductive age is reported to be 10% (Houston 1984; Olive and Schwartz 1993), and no population-based prevalence data exist for PCOS.

With respect to developmental outcomes, population-based data are available regarding the prevalence of birth defects. Approximately 2-3% of infants are born with major birth defects (Holmes 1997). The full impact of prenatal testing on the prevalence of birth defects has not been delineated. The prevalence of birth defects increases (approximately 5%) when all defects (i.e., major and minor) are included. However, identifying the prevalence of minor defects is problematic, given differences in clinical assessment, recognition and reporting of defects, and variations across state-birth-defects registries in the recording of minor defects. Other developmental outcomes that need to be considered in assessing developmental toxicity include fetal and infant growth and developmental disabilities during infancy and childhood. Boyle et al. (1994), citing data from the 1988 National

Health Interview Survey, found that 17% of children in the United States were reported to have a developmental disability. The prevalence rates for specific developmental disabilities per thousand 10-year-old children were 10.3 for mental retardation, 2.0 for cerebral palsy, 1.0 for hearing impairments, and 0.6 for visual impairments. Another study found that about 4% of U.S. children aged 17 years or younger are reported by their parents to have delays in growth and development, and that 6.5% of children have learning disabilities (Zill and Schoenborn 1990).

The causes of many developmental abnormalities are unknown. As an example, Nelson and Holmes (1989), through a careful evaluation of approximately 70,000 children and their families, were able to account for 57% of birth defects: mutations (28%), multifactorial conditions (23%), uterine factors and twinning (3%), and exposure to chemical and physical agents found in the environment (3%). Thus, the etiology of at least 43% of all birth defects could not be determined. Prevention of developmental defects depends on understanding their causes and is important in reducing the tremendous societal and financial burden.

Economic costs for adverse reproductive and developmental outcomes are noteworthy and expected to grow. Annual costs for infertility treatment in the United States exceed one billion dollars (U.S. Congress 1988). According to the Centers for Disease Control and Prevention (CDC), the cost to society for developmental defects is massive (i.e., the lifetime costs for children born annually with 17 of the most common birth defects and cerebral palsy is over $8 billion (CDC 1995)). However, these abnormalities affect only 22% of children with birth defects, and the cost estimate does not consider costs associated with many other developmental disorders. A recent study estimated the total lifetime costs for persons born in 1996 with mental retardation, autism, or cerebral palsy to be $47 billion, $4.9 billion, and $12 billion, respectively (Honeycutt et al. 1999).

To reduce the number of reproductive and developmental disorders caused by exposure to chemical and physical agents, reproductive and developmental toxicity testing generally is conducted in laboratory animals because at present no other approach is considered predictive of reproductive and developmental effects and, also, data in humans are often not available. Several regulatory agencies and other organizations have developed evaluative guidelines and processes for identifying and assessing reproductive and developmental toxicity (U.S. Environmental Protection Agency (EPA) 1991, 1996a; European Commission (EC) 1992; J.A. Moore et al. 1995a; California Environmental Protection Agency 1991). Each agency or organization has established criteria for evaluating data on the reproductive and developmental effects of exposures to agents, as summarized in Table 1-1. The International Programme on Chemical Safety (IPCS) (1994) identified current differences in risk assessment procedures for reproductive and developmental toxicity among several countries. IPCS also has proposed ways to improve the international harmonization of those procedures.

Concern regarding reproductive and developmental hazards in the workplace, including military facilities, has increased significantly in recent years. In 1997, Congress passed a law, as part of the National Defense Authorization Act, concerning health care coverage for children with medical conditions caused by parental exposure to hazardous materials while serving as members of the Armed Services (Public Law 104-201, Section 704). The law states, in part, that a plan would be developed for ensuring the provision of medical care to any natural child of a member of the Armed Forces who has a congenital defect or catastrophic illness, proven to a reasonable degree of scientific certainty on the basis of scientific research to have resulted from exposure of the member to a chemical warfare agent or other hazardous material to which the member was exposed during active military service. The Department of Defense is required to develop a plan for compliance.

With respect to health surveillance for deployed forces, The National Academies Institute of Medicine recommends the development of strategies for the protection of reproductive health in men and women and fetal development and well-being of offspring (IOM 1999). In sum, assessment of reproductive and developmental toxicity requires consideration of a wide spectrum of possible endpoints in both men and women and their offspring. Appreciation of the subtle and methodological nuances underscoring successful human reproduction need to be appreciated in the systematic evaluation for assessing reproductive and developmental toxicity.

TABLE 1-1 Comparison of the Criteria for Evaluating the Data on the Reproductive and Developmental Effects of Agent Exposures

Document (in chronological order)	Intended Purpose	Definitions	Criteria
EPA – Developmental Toxicity Risk Assessment Guidelines (EPA 1991)	Provide guidance on the evaluation of human and animal data on developmental toxicity. Guidelines describe the procedures EPA follows in evaluating potential developmental toxicity by analyzing and organizing data for risk assessments.	Developmental toxicology – The study of adverse effects on the developing organism that may result from exposure prior to conception (either parent), during prenatal development, or postnatally to the time of sexual maturation. Adverse developmental effects may be detected at any point in the lifespan of the organism. The major manifestations of developmental toxicity include death of the developing organism, structural abnormality, altered growth, and functional deficiency.	*Sufficient versus insufficient evidence.* Sufficient human evidence includes data from epidemiological studies that provide convincing evidence that a causal relationship is or is not supported. A case series with strong supporting evidence may also be used. Sufficient experimental animal evidence-limited human data includes data from experimental animal studies and/or limited human data that provide convincing evidence that the potential for developmental toxicity exists. Insufficient evidence refers to a data set that provides less than the minimum sufficient evidence necessary for assessing the potential for developmental toxicity (e.g., there are no developmental toxicity data available or there are methodological limits of studies).

TABLE 1-1 (*Continued*)

Document (in chronological order)	Intended Purpose	Definitions	Criteria
California Proposition 65 (1991)	Protect people of California from reproductive (including developmental) toxicants to ensure their health and well-being.	Reproductive toxicity includes developmental toxicity, female reproductive toxicity, and male reproductive toxicity. Developmental toxicity includes adverse effects on the products of conception.	The Safe Drinking Water and Toxic Enforcement Act of 1986 (Proposition 65) provides three mechanisms for listing a chemical as a reproductive toxicant: (1) if an authoritative body (e.g., EPA, National Institute for Occupational Safety and Health recognized by the DART[a] scientific advisory board identifies a chemical as such, (2) if a state or federal government formally requires labeling as a reproductive toxicant, and (3) a de novo listing based on the opinion of the DART scientific advisory board using a "weight-of-evidence" approach. The total body of evidence is categorized as sufficient, limited, deficient, and null.
European Commission (EC 1992)	Develop a system of classification for substances that might be considered toxic to reproduction.	Reproductive toxicity includes impairment of male and/or female fertility and effects on development of the progeny (developmental toxicity). Developmental toxicity includes nonheritable adverse effects on the further development of the offspring up	Agents that cause adverse effects on reproduction are classified as those that affect fertility and those that affect offspring. Each classification is subdivided into three categories: (1) agents known to impair fertility or development in humans, (2) agents that should be regarded as though they impair fertility or develop-

	to attainment of sexual maturity and adult life.		ment in humans, and (3) agents that cause concern for humans because of possible fertility or developmental toxic effects. Evidence required to place an agent in each category is listed. If an agent does not meet criteria of a category, it should not be classified.
International Programme on Chemical Safety (IPCS 1994)	Harmonization of risk assessment for reproductive and developmental toxicity by establishing a glossary for terminology, identifying minimum database requirements, identifying areas of difference in risk assessment procedures and approaches for reconciliation; and consider harmonized formats for data reporting.	Merged EC and EPA definitions. Reproductive toxicity includes adverse effects on sexual function and fertility in males and females as well as developmental toxicity. The toxicity might be expressed as alterations to the female or male reproductive organs, the related endocrine system, or pregnancy outcomes. Developmental toxicity in its widest sense includes any effect interfering with normal development before or after birth. The occurrence of adverse effects on the developing organism might result from exposure before conception (either parent), during prenatal development, or postnatally to the time of sexual maturation. Adverse developmental effects may be detected at any point in the lifespan of the organism.	Establishes criteria for low to high concern using a minimum data set. A standardized format for the summary would comment on the sufficiency and quality of data to draw conclusions.

TABLE 1-1 (*Continued*)

Document (in chronological order)	Intended Purpose	Definitions	Criteria
Evaluative Process (J.A. Moore et al. 1995a)	To develop an evaluative process for assessing reproductive and developmental toxicity.	Developmental toxicity consists of adverse effects on the developing organism that might result from exposure before conception (either parent), exposure during prenatal development, or exposure during postnatal development from birth to sexual maturation. Reproductive toxicity can involve alterations in the reproductive organs or in the related endocrine systems resulting in a variety of effects among men and women.	*Sufficient versus insufficient evidence.* Three generic criteria are applied to describe data that are insufficient: (1) there are no data, (2) deficiency in study design or insufficient detail available to allow analysis, and (3) data are insufficient to reach a definitive conclusion. The evaluative process requires a *weight of evidence* approach to determine (in)sufficiency of data as summarized in a narrative document. The evaluation process determines whether the collective toxicity data are (in)sufficient to judge that there is an adverse effect under specified exposure conditions or whether the data are (in)sufficient to conclude the absence of adverse effects under specified exposure condition.
EPA – Reproductive Toxicity Risk Assessment Guidelines (EPA 1996a)	Provide guidance for assessing the effects of environmental agents that might adversely affect human health, including the reproductive system.	Reproductive toxicity is the occurrence of biologically adverse effects on the reproductive systems of females or males that might result from exposure to harmful substances in the environment. The toxicity might be manifested	*Sufficient versus insufficient evidence.* Sufficient evidence includes data that collectively provide enough information to judge whether or not a reproductive hazard exists within the context of effect as well as dose, duration, timing, and route of exposure. Both human and animal evi-

in various ways. Developmental toxicity is the occurrence of adverse effects on the developing organisms that might result from exposure prior to conception (either parent), during prenatal development, or postnatally to the time of sexual maturation. These adverse effects can be manifested in various ways over the lifespan of the organism.

dence may be included. Insufficient evidence refers to data less than the minimum for sufficient evidence of reproductive toxicity, largely reflecting methodological limits of reported studies or the absence of studies.

[a]DART refers to reproductive and developmental toxicity.

SUBCOMMITTEE'S TASK

Every year, the U.S. Navy screens hundreds of chemical substances for potential toxicity to determine whether they can be used safely in the workplace. Although the Navy reviews available data on the reproductive and developmental toxicity potential of those agents, its health hazard evaluation process is not currently designed to emphasize assessment of reproductive and developmental effects.

Because the Navy wishes to protect its male and female military and civilian personnel from reproductive and developmental hazards, it seeks to incorporate a formalized, state-of-the-art process for identifying hazards in its current health hazard evaluation process. Therefore, the Navy requested that the National Research Council (NRC) recommend an approach that can be used to evaluate agents for potential reproductive and developmental toxicity. The NRC assigned this project to the Committee on Toxicology, which convened the Subcommittee on Reproductive and Developmental Toxicology. The subcommittee was charged with the following tasks:

- Develop a process for assessing the reproductive and developmental toxicity potential from exposures to chemicals and physical agents.
- Develop a strategy for dealing with the potential reproductive and developmental toxicity of exposures to chemicals and physical agents for which little or no information is available.
- Conduct pilot demonstrations on two chemicals using the process developed by the subcommittee.
- Identify reliable sources for assessment of reproductive and developmental toxicity.
- Identify areas of needed research.

For the purposes of this study, the primary focus is on avoiding occupational exposures to agents that potentially cause reproductive and developmental toxicity. Therefore, this report recommends a process that focuses on evaluating exposures to adult male and female military and civilian personnel and on predicting the effects of those exposures on those adults and their children. It does not specifically address direct exposures to children of military and civilian personnel, including children living on military bases.

Agents that have the potential to cause reproductive and developmental toxicity should be substituted with less hazardous materials or surveillance should be increased in an effort to control exposures. The Subcommittee on Reproductive and Developmental Toxicology believes it is inadvisable to designate agents as toxic or nontoxic; rather, the agent and the risk of adverse reproductive or developmental effects should be considered only in the context of exposure. Assessment of exposure accounts for both the agent itself and for the conditions of exposure, including the amount, route, timing, duration, and pattern of exposure.

The subcommittee recognizes the need for the Navy to use a screening process in which decisions are made in a dichotomous manner (to use or not to use a particular agent). Such decisions can be made by considering the exposure scenario that is anticipated in the workplace. In this report, the subcommittee describes a process by which an exposure can be estimated that is unlikely to be associated with reproductive and developmental toxicity. If the Navy workplace scenario is anticipated to result in human exposures appreciably lower than that estimate, then for policy decisions, the exposure can be regarded as acceptable. If the anticipated human exposure is higher than the estimated nontoxic level, the use of the agent in question can be regarded as unacceptable, and alternative agents can be evaluated or exposure control measures (e.g., use of masks and protective clothing) can be used.

Data on the reproductive and developmental effects of the agents of interest often are sparse. When they are available, the quality of the studies from which the information is obtained can be highly variable. The quality and quantity of the data are related to the level of confidence in assessing concern about exposure to an agent. When adequate data sets are available, there is a high level of confidence in determining a low (if no significant toxicity is expected at anticipated exposures) or high (if toxicity is expected at anticipated exposures) degree of concern. When the data set is inadequate, the level of confidence in assessing the degree of concern in lessened.

The process described by the subcommittee requires the exercise of considerable judgment, brought to bear in assessing the adequacy of data for estimating potential reproductive and developmental toxicity of agents under specific conditions of exposure. Once there has been a determination of the exposure at which adverse effects are unlikely, judgment is required in the evaluation of other characteristics of the agent or exposure conditions that might make it advisable to alter the exposure estimate for a given workplace scenario. *Therefore, the subcommittee believes the process should be implemented by a team of scientists who have training and experience in assessing reproductive and developmental toxicity.*

Reproductive and developmental toxicity often is equated with pregnancy effects (maternal and developmental effects) of exposures of interest. The subcommittee believes that pregnancy effects of exposures are important, but attention also should be directed to the potential for reproductive toxicity in males and nonpregnant females, and to the potential for paternally mediated adverse effects of exposures. Although examples of the latter toxicity in humans are not readily available, several agents have been shown to have such effects in experimental animals; therefore, it is important to take seriously any data that suggest a paternally mediated effect. Reproductive systems of elderly adults also could be susceptible to adverse effects. In such cases, the concern is not with procreative competence but with biological function maintained by the gonads (e.g., hormone production) and with reproductive senescence. Any data that suggest such an effect should be considered. *The subcommittee recommends that an assessment not be considered complete unless it includes consideration of potential adverse reproductive and developmental consequences of exposure of both the male and the female. The absence or inadequacy of data on one or more of the components of reproductive toxicity (i.e., male reproductive effects, female reproductive effects, or developmental effects) does not equate with lack of effect.*

ORGANIZATION OF THE REPORT

In addition to this Introduction, this report is organized into four chapters. Chapters 2 and 3 outline the process the subcommittee recommends for evaluating the potential reproductive and developmental effects of exposures to agents. Chapter 2 outlines the principles for evaluating reproductive and developmental toxicity data. The product of that evaluative process is typically an exposure level—called the unlikely effect level (UEL)—that is assumed not to pose appreciable risk of reproductive and developmental effects, adjusted for the exposure scenario of concern. The UEL is similar to the EPA's acute and chronic Reference Doses (RfD) and the U.S. Food and Drug Administration's Acceptable Daily Intake (ADI) except that it is specific for reproductive and developmental effects and is derived specifically for the exposure duration of concern in humans. Chapter 3 discusses how the UEL is derived in the evaluative process and how it can be compared with an anticipated human exposure such as found in the workplace. Chapter 4 sets forth a strategy for evaluating exposures for which there are few or no data on reproductive or developmental toxicity. The subcommittee's recommendations are presented in Chapter 5. This report also contains four appendixes: Appendix A contains examples of the application of the proposed evaluative process to two specific chemical agents, Appendix B describes and evaluates various sources of information on reproductive and developmental toxicity, Appendix C describes and evaluates human study designs, and Appendix D describes experimental animal study designs and discusses qualities and limitations for each type of study.

2

The Evaluative Process:
Part I. Assessing the Available Data

This chapter and the next are based on a paper, "An evaluative process for assessing human reproductive and developmental toxicity of agents," published in *Reproductive Toxicology* (Moore et al. 1995a) that calls for the systematic application of knowledge and judgment to assess agents for reproductive and developmental toxicity in a practical, open, and informative manner. In addition, the U.S. Environmental Protection Agency's (EPA) guidelines for developmental toxicity (1991) and reproductive toxicity (1996a) risk assessment, and several additional sources reviewed in Chapter 1, were used extensively. Several principles and objectives that are incorporated in the evaluative process are described below. The principles and objectives are followed by details of the evaluative process. The description of the evaluative process continues in the next chapter. In that chapter, the interpretation of toxicity data, integration of toxicity and exposure data, and quantitative assessment steps are covered.

PRINCIPLES AND OBJECTIVES

Use of Data and Judgment

The evaluative process uses both scientific data and scientific judgment. The data required to identify an agent as toxic should adequately demonstrate adverse effect and dose-response relationships for general toxicological responses and for reproductive and developmental effects. Furthermore, there is a significant need for data that characterize human exposure or provide reasonable estimates based on the pattern of use of the agent. The essence of the evaluative process is that the interpretation of those data should reflect expert judgment, rather than acquiesce to the passive use of a repetitive series of default assumptions. A valuable adjunct to the evaluation of an exposure to an agent is the inclusion of a statement of what is known and the certainty with which it is known. That should lead to the identification of critical data needs that might stimulate investigations to yield useful information that will enhance certainty of judgment and better serve the U.S. Navy.

Weight of Evidence

With a weight-of-evidence approach that considers both toxicity and human exposure information, evaluators can determine whether human or experimental animal data can reasonably be used to predict reproductive or developmental effects in humans under particular conditions of exposure. The approach must distinguish those agents for which there is firm evidence about human risk potential, based on relevant data, from those for which the potential for human effects is uncertain or unlikely. It will aid in setting priorities and developing programs to protect personnel from undue exposure to toxic quantities of agents or from undue costs of unnecessary control measures.

Using a weight-of-evidence approach to communicate a judgment about human risk, taking into consideration exposure potential, should diminish reliance on the assumption that reproductive and developmental toxicity observed in animals predicts similar effects in humans. Because the evaluative process requires a judgment about human risk potential based on the weight of the evidence, its approach and its results will be useful to the Navy. That approach differs from several programs that assess carcinogenic potential, including the International Agency for Research on Cancer (IARC) monographs, which invoke "sufficiency of evidence" determinations for experimental data; the Science Advisory Panel for the California Proposition 65 listing process, which follows a similar procedure in its review of carcinogenicity data; and the report on carcinogens produced by the National Toxicology Program (NTP), which primarily lists the results of experimental animal studies. IARC and NTP clearly state that their deliberations do not represent a complete assessment of human risk potential, but their monographs and lists continue to be misused for that purpose.

Threshold Assumption

Use of a threshold assumption for low dose-response relationships implies that there is an exposure level below which an adverse effect is not expected to occur. The assumption of a threshold has been made historically for the chemical induction of many types of reproductive and developmental effects as well as for other noncancer health effects. This is in contrast to the case for carcinogens which historically have been assumed to have no threshold. Recent emphasis on using mechanistic or mode-of-action information to improve the risk assessment process (EPA 1996b) and to harmonize the approaches used for cancer and other types of health effects (Bogdanffy et al. in press), underscores the use of mechanistic information in the weight-of-evidence approach for low-dose extrapolation. For example, some nongenotoxic carcinogens may not have a linear dose-response relationship at low doses (Andersen et al. 2000), whereas some agents that produce reproductive and developmental toxicity may act through a genetic mechanism or an endogenous mechanism that is additive to background, and therefore be more likely to exhibit a linear dose-response relationship at low doses (Gaylor et al. 1988). These types of mechanisms tend to blur the distinction between the default use of a linear low-dose extrapolation for cancer and a threshold assumption for other health effects which defaults to the application of uncertainty factors to the no-observed-adverse-effect level (NOAEL), lowest-observed-adverse-effect level (LOAEL), or benchmark dose. Consideration of mechanistic information (i.e., toxicokinetic and toxicodynamic data) should be a major factor in the weight-of-evidence process and in deciding how to proceed with low-dose extrapolation. In the absence of any mechanistic or mode-of-action information, the default assumption continues to be a threshold or low dose nonlinear dose-response relationship for health effects other than cancer, but this assumption should continue to be explored through the development and application of mechanistic data in the risk assessment process.

Narrative Statement

Communicating the results of a weight-of-evidence evaluation is best accomplished through a narrative document. A narrative permits expression of the degree of certainty associated with a judgment about the scientific evidence. The document must use terms that are meaningful to Navy policy officials or decision makers, it must define those terms carefully, and it must use them consistently. The narrative must be clear in explaining the basis of the judgment, the breadth of expert support, the degree to which the judgment reflects the actual information, and the assumptions made in the absence of information.

Certainty

Documents produced under the evaluative process will clearly describe the level of confidence in the evaluative judgment. The need to invoke a series of default assumptions will signify progressively greater degrees of uncertainty. Certainty based on the interpretation of essential data should be distinguished from "certainty based on defaults," where default assumptions force evaluators to designate an agent as having toxic potential. Conservative default assumptions, based on prudent public health concerns, have a rightful place in the options available to risk assessors and managers. Such assumptions should be used only where absolutely necessary, and always openly.

Finally, because the evaluative process adopts an open, candid, narrative form of communication, it minimizes the dissemination of inappropriate or simplistic statements that are commonly misused and that are needlessly alarming.

Use All Relevant, Acceptable Data

Reaching a determination about an agent's potential toxicity to humans is best done by a consideration of all relevant experimental animal and human data. Decisions to use either published or unpublished data should depend on the quality and completeness of the data set. Unfortunately, publication in the open scientific literature does not in itself qualify data as acceptable for evaluation. Many published articles present data in insufficient detail to allow them to be of use in risk evaluations.

Whether data are judged acceptable from the perspective of sound scientific design and interpretation will depend heavily on the actual review of specific studies. Good laboratory practices have been promulgated by the Organization for Economic Cooperation and Development (OECD 1987), the U.S. Food and Drug Administration (FDA 1987), and the U.S. Environmental Protection Agency (EPA 1990). Good laboratory practices can serve as useful guides for assessing the quality and completeness of reported data. Comparing the test design and completeness of data reporting to what is outlined in test guidelines and procedures might be of particular value. Other factors that should be considered include statistical power, analytical approaches, data presentation, and consistency with other results.

Qualities and Limitations of
Reproductive and Developmental Toxicity Studies

Developmental toxicity studies typically assess whether structural abnormalities are associated with administration of an agent to a pregnant female during major organogenesis in the developing embryo. General reproductive effects can be assessed through analysis of various types of mating studies. Reproductive and developmental toxicity studies provide much useful information on a chemical's potential to cause adverse reproductive and developmental effects. However, the limitations of each study type should be recognized. For example, the prenatal developmental toxicity study provides information on the effects of repeated exposure to an agent during the period of major organogenesis, but does not follow animals postnatally to evaluate aspects such as reversibility and repair or organ function. Detailed descriptions of various study types and their qualities and limitations are presented in Appendix D.

Characterizing Data

The evaluative process uses three generic criteria for judging data insufficient:

- There are no data.
- The studies are of limited utility as a result of deficiencies in design.
  or execution, or because the data are insufficiently detailed to allow an independent analysis.
- The available studies are acceptable, but the data are insufficient to reach a definitive conclusion because they do not span a sufficient number of outcomes; the study might, however, offer useful supplemental information.

Data sets that are insufficient for evaluating reproductive or developmental toxicity do not arise solely from studies that are unreliable and therefore unworthy of consideration. Information from in vitro or nontraditional in vivo studies, for example, frequently provides enough experimental evidence to corroborate other evidence for an adverse effect. Alone, however, those studies might not provide enough evidence to be considered sufficient to identify an adverse effect.

A judgment that data are insufficient to establish an adverse effect does not mean that they are sufficient to establish *lack* of an adverse effect. Such a presumption would be erroneous. Sufficiency is a designation with stringent criteria. The criteria required for a sufficient data set are discussed later.

Expert Review Team

Evaluations of exposures to agents should involve a group of experts. The breadth of expertise required is rarely found in one person and group review ensures that the views held by each member are subjected to the scrutiny and acceptance of scientific peers. The group should include epidemiologists and experts in toxicology and related areas (e.g., reproductive toxicologists, developmental toxicologists, developmental neurotoxicologists, risk assessors, biostatisticians), as well as in human exposure to the chemicals of interest. A rotating core of scientific members who serve for fixed periods on a series of working groups will enhance consistency of reviews. This model is in use by the NTP's Center for the Evaluation of Risks to Human Reproduction, which was established in 1998. Using the evaluative process is resource- and time-intensive and, therefore, in some cases, the Navy may want to consult existing sources of information which provide detailed evaluations developed by experts in reproductive and developmental toxicology. The detailed evaluations that are available as well as other sources of information are described in Appendix B.

GENERAL DESCRIPTION

The evaluative process recommended by the subcommittee (based on the process described by Moore et al. 1995a) outlines a systematic, sequenced procedure for reviewing data on animal and human reproductive and developmental toxicity, on general toxicological and biological parameters, and on the conditions of use that result in human exposure. The goal is to determine whether exposure to an agent could cause reproductive or developmental toxicity in humans. Expert judgment is applied in a series of steps that reflect the systematic thought sequences used by most experienced risk assessors. Brief summaries that describe each step appear below, followed by more detailed presentations in the rest of this chapter and in Chapter 3.

The section on *exposure data* discusses the pattern and degree of human exposure to the agent. It primarily considers occupational exposures and develops numerical estimates of exposure from what is known about those uses and exposures.

The section on *general toxicological and biological parameters* reviews and summarizes chemical data and basic toxicity information available on the agent of interest and reviews data on absorption, distribution, metabolism, and excretion in humans and experimental animals.

The section on *developmental and reproductive toxicity* reviews data from human and animal studies. To ensure adequate assessments of both types of data, experts review each type of data independently and prepare synopses of individual studies.

In the step for *integration of toxicity and exposure information,* the existing data on human and experimental animal developmental and reproductive toxicity are evaluated together for evidence of complementarity or inconsistency. Those evaluations are then assessed in terms of the known data on basic toxicity and pharmacokinetics. The result is an integrated judgment about the relevance of all the data for predicting human risk. If the expert committee members judge that the toxicity data are relevant to humans, the committee undertakes a *quantitative evaluation.* Finally, the toxicity and exposure data are integrated to characterize risk.

When the data reviewed are deficient, the ensuing judgments usually involve a large degree of uncertainty. The identification of *critical data needs* provides a focus on research that can materially enhance the certainty of future judgments about the agent's potential risk.

A *summary* reviews the scientific judgments and conclusions formed in the steps above and conveys the level of confidence in the judgment. The summary is written in a narrative style which the subcommittee considers to be the best way to present such information to Navy environmental health professionals. The narrative is central to accurate interpretation of the scientific judgments and conclusions about the exposure of interest. Agents present a reproductive or developmental risk to human health only under certain conditions. Single-letter or word designations, such as "positive" or "negative," or labeling a chemical as a "reproductive toxicant" cannot effectively communicate that critical fact. Nor can essential facts about such parameters as frequency, duration, and route of exposure, susceptible populations, age, and reproductive status be conveyed without some sense of context. For those reasons, the narrative form is crucial.

The last step is a listing of references for papers and studies of the agent of interest.

DETAILS OF THE EVALUATIVE PROCESS

The sections below detail the steps of the evaluative process recommended by the subcommittee. Box 2-1 is a sample table of contents from an evaluation of lithium (Moore et al. 1995b) using a similar process.

Box 2-1 Example Table of Contents — Assessment of Lithium

INTRODUCTION

1. **Exposure Data**
   1.1 Consumer Exposure
   1.2 Environmental Exposure
   1.3 Occupational Exposure
   1.4 Exposure Estimates
2. **General Toxicological and Biological Parameters**
   2.1 Chemistry
   2.2 Basic Toxicity
   2.3 Pharmacokinetics
3. **Reproductive and Developmental Toxicity Data**
   3.1 Human Data
      3.1.1 Developmental Toxicity
         3.1.1.1 Register Studies
         3.1.1.2 Prospective Studies
         3.1.1.3 Retrospective Studies
         3.1.1.4 Clinical Case Reports
      3.1.2 Reproductive Toxicity
         3.1.2.1 Developmental Toxicity
         3.1.2.2 Reproductive Toxicity
   3.2 Experimental Animal Toxicity
      3.2.1 Developmental Toxicity
         3.2.1.1 Studies in Mice
         3.2.1.2 Studies in Rats
         3.2.1.3 Studies in Rabbits, Monkeys, and Pigs
      3.2.2 Reproductive Toxicity
         3.2.2.1 Female Reproductive Toxicity
         3.2.2.2 Male Reproductive Toxicity

4. **Integration of Toxicity and Exposure Information**
   4.1 Interpretation of Toxicity Data
      4.1.1 General Toxicity and Pharmacokinetics Conclusions
      4.1.2 Developmental Toxicity
         4.1.2.1 Conclusions
      4.1.3 Reproductive Toxicity
         4.1.3.1 Female Reproductive Toxicity
         4.1.3.2 Male Reproductive Toxicity
         4.1.3.3 Conclusions
   4.2 Default Assumptions
   4.3 Quantitative Evaluation
      4.3.1 Developmental Toxicity
      4.3.2 Reproductive Toxicity
5. **Critical Data Needs**
   5.1 Developmental Toxicity
   5.2 Female Reproductive Toxicity
   5.3 Male Reproductive Toxicity
6. **Summary**
   6.1 Background
   6.2 Human Exposure
   6.3 Toxicology
      6.3.1 Developmental Toxicity
      6.3.2 Reproductive Toxicity
         6.3.2.1 Female Reproductive Toxicity
         6.3.2.2 Male Reproductive Toxicity
   6.4 Quantitative Evaluation
      6.4.1 Developmental Toxicity
      6.4.2 Reproductive Toxicity
   6.5 Certainty of Judgments and Data Needs
      6.5.1 Developmental Toxicity
      6.5.2 Reproductive Toxicity
7. **References**

Source: Moore et al. (1995b).

Exposure Data

Human exposure data are evaluated to achieve three goals:

- To identify potentially exposed populations.
- To identify potential pathways of exposure and to describe the parameters associated with each pattern of use, including route, dose, duration, frequency, timing, age, and number of people potentially exposed.
- To estimate the range of exposure and thus obtain quantitative estimates of exposures associated with each pattern of use.

Although human exposure data are essential for accurate evaluation of an agent's risk potential, data of sufficient quality and quantity are frequently unavailable. Thus, there is uncertainty in the exposure component of the evaluative process, even as there is in hazard characterization. When toxicity data indicate the potential for an adverse effect, the need to estimate the nature of human exposure becomes imperative. In those instances, exposure estimates can be derived using modeling approaches based on data from other sources, and one or more default assumptions can be used. The greater the number of default assumptions employed, the greater the uncertainty about the accuracy of the expert judgment.

A chemical might have a variety of uses, and the concentration, route, and frequency of exposure can differ for each use. The physical form of the chemical and the presence of other agents also might vary with use. Those factors can dramatically influence both the probability that exposure will lead to absorption into the body and the rate at which absorption occurs. Some uses might lead to indirect exposures, perhaps resulting from deliberate, incidental, or accidental environmental releases of the chemical. Pesticide residues in food are an example of exposure that arises from a deliberate environmental release. Incidental or deliberate releases of pesticides, through normal use, might lead to exposure through drinking water or in respired air. Some exposures are direct: Examples include consuming a chemical as a drug or using chemicals to mask odors. Although the frequency and intensity of exposure to an agent are typically greatest in occupational settings, sometimes use of certain products by Navy personnel outside the workplace can lead to episodes of exposure intensity that approach or exceed occupational exposures. Examples include the use of cosmetics and nonprescription drugs; pesticide applications in the home; furniture refinishing; and home remodeling.

Because it is frequently difficult to establish directly many patterns of agent use, data on exposure can be estimated by indirect modeling (EPA 1992). Indirect assessments use available information on concentrations of chemicals in exposure media, and information about when, where, and how individuals might have contact with the exposure media. Models and a series of exposure factors (e.g., agent concentration, contact duration, contact frequency) are then used to estimate exposure. The models can be deterministic or probabilistic. A deterministic model provides a point estimate of exposure; a probabilistic model considers the range of estimates and provides a probability distribution of exposures. Data sets are rarely complete and, therefore, exposure estimates are developed using various default assumptions (combined with the modeling estimates). For example, to estimate the risk posed by pesticides in foods, EPA initially assumes that residues are at tolerance levels and that 100% of a crop has been treated (EPA 1999). To maintain occupational exposure limits, personal exposure monitoring techniques can be used.

Exposure assessments generally focus on a single chemical and a single route of exposure. However, there have been recent efforts to examine multiple pathways of exposure. The current approach is to add the single point estimates for each exposure source to arrive at a sum. Research continues on developing new data and exposure models for estimating multiple-pathway exposures.

It might not be necessary to review each exposure parameter on a chemical-and use-specific basis. Exposure paradigms and values that are in regular use in government agencies or that are recommended by scientific organizations could be adopted. The American Conference of Government Industrial Hygienists' (ACGIH) Threshold Limit Values (TLVs) and Biological Exposure Indices (BEIs) (ACGIH 2000), the U.S. National Institute for Occupational Safety and Health's (NIOSH) recommended exposure limits (RELs) (NIOSH 2000), and the U.S. Occupational Safety and Health Administration's (OSHA) permissible exposure limits (PELs) and short-term exposure limits (STELs) (29

CFR Part 1910 Subpart Z) provide guidance on controlling workplace exposures. The specific paradigm and values should be referenced in each instance. Consistent with the preference to supplant default assumptions with actual data, the process should make reasonable efforts to ascertain the availability and quality of such data and explicitly state where the process makes use of default assumptions or actual data.

Several exposure conditions are unique for reproductive and developmental toxicity. For example, an adverse effect on reproductive function could depend on when exposure occurs during male and female development. Exposure that occurs during prenatal and postnatal development can affect reproductive function differently than exposure that occurs in adulthood. Different groups (embryo, fetus, neonate, juvenile, young adult, and older adult) can vary in susceptibility.

For male and female reproductive toxicity, exposure assessments should consider the duration and period of exposure during development (prenatal, prepubescent, and reproductive) and physiological state in females (pregnancy, lactation, peri- and postmenopausal).

Exposure of the conceptus to a chemical agent depends on maternal absorption, distribution, metabolism, and placental metabolism and transfer. Transit time in the conceptus also depends on its ability to metabolize and excrete the chemical. In a few cases, a chemical agent will have its primary effect on the maternal system (Daston et al. 1994), and the effect on the conceptus will not depend on exposure to the agent and its metabolites but on some factor induced in the mother. Infants can be exposed via breast milk, especially to metals and fat soluble-agents. There are data that suggest that lead is mobilized during pregnancy and lactation, along with calcium from bone stores, and excreted in milk in women during lactation (Silbergeld et al. 1988; Silbergeld 1991; Gulson et al. 1997). Thus, exposure of the conceptus and child often will not be the same as for the pregnant or lactating mother, and measurement of the agent in cord blood and in breast milk could give a better exposure estimate.

Exposures during pregnancy should be characterized as to the time during pregnancy, at least during the first, second, or third trimester (although identification of the week or month is more desirable). Sensitivity to a particular agent can differ at various times during pregnancy, and exposure estimates should be characterized for as many effects as possible. In addition, exposure to either parent prior to conception should be considered as a possibility in the production of adverse developmental effects (Olshan et al. 1994).

Even a single exposure, no matter the duration, might result in reproductive or developmental toxicity. Agents that show accumulation with repeated exposures or that have a long half-life will result in greater exposures over time. The pattern of exposure is extremely important in predicting outcome, and it is usually difficult to extrapolate results from one pattern to another unless pharmacokinetic data are available to illuminate the differences. In the case of intermittent exposures, peak exposures and averages over time should be considered.

Unique to exposure during development is that there are often latent effects, except in cases of spontaneous abortion that occur during or shortly after exposure. For example, effects of exposure during pregnancy might not be manifested until after birth in the form of structural malformations, impaired growth, cancer, mental retardation, or other functional defects. In some cases, the results of exposure might not become apparent until long after the developmental period, including neurotoxic effects that are not evident until adulthood or until another factor (e.g., disease, other environmental challenges, aging) intervenes (Barone 1995).

Each evaluation should describe patterns of use, with specific emphasis on the potential for direct or indirect exposure. Data on real or estimated levels of exposures should be collected, as should information on population distributions, intensity, routes, timing, and durations. Other data of value include industrial hygiene measurements at the point of manufacture, materials balance (input, products, waste) at the point of production, shipping patterns and methods of transportation, industrial hygiene measurements at point of use if relevant, and ambient-air monitoring data. Where there are multiple patterns of use or routes of exposure, an effort should be made to determine whether some patterns account for larger exposures than others. The evaluations also should consider data on environmental fate and transport. Toxic release inventory data and data residues in food and potable water should be included, if available. Evaluations of drugs should gather basic dosimetry information, including a profile of the user population and pharmacokinetic data.

Valuable general references include the *Exposure Factors Handbook* (EPA 1997a) and EPA's exposure assessment guidelines (EPA 1992). The Pesticide Assessment Guidelines, Subdivision U, Applicator Exposure Monitoring (EPA 1987) and the Standard Operating Procedures (SOPs) for Residential Exposure Assessments (EPA 1997b) should be consulted in the analysis of specific occupational and residential exposures. Guidance on controlling workplace exposures to chemicals and physical agents can be obtained from ACGIH's TLV occupational exposure guidelines and BEIs (ACGIH 2000), the NIOSH Pocket Guide to Chemical Hazards (NPG) (NIOSH 2000), and the Occupational Safety and Health Guidelines for Chemical Hazards (29 CFR Part 1910 Subpart Z).

General Toxicological and Biological Parameters

Chemistry

Generic chemical class data are often relevant to assessing potential toxicity and should be a part of any evaluation. The relevant information includes structure-activity relationships and physical-chemical properties, such as melting point, boiling point, solubility, and octanol-water partition coefficient. Physical-chemical properties affect an agent's absorption, tissue distribution, biotransformation, and degradation in the body.

Basic Toxicity

Reproductive or developmental toxicity endpoints must be interpreted in the context of general toxicity that could also occur in the same animals. Toxic effects reported from other studies can be particularly valuable because excessive toxicity could significantly confound the interpretation of a reproductive or developmental toxicity study. Observations from studies of other toxicity endpoints might either strengthen or weaken the conclusions to be drawn from a reproductive or developmental study and provide information about target organs that should be evaluated further in developing animals.

Relevant toxicity data typically originate from acute (single dose) or repeated-dose studies of up to 90-days' duration. Protocols for those studies have been developed by FDA, EPA, OECD and other federal regulatory agencies. International entities promulgate broadly accepted test guidelines.

Acute Studies

Acute studies, for which the primary endpoint is lethality, are most often conducted in rats or mice by the oral or inhalation routes, or in rabbits by the oral or dermal routes of exposure.

Although acute lethality data are not predictive of reproductive or developmental toxicity, they are useful indicators of divergences in species or route sensitivity. When there are significant species differences in acute toxicity, for example, one would also expect differences between species in the doses that would cause reproductive or developmental toxicity. Doses for such studies are selected on the basis of general toxicity parameters, such as mortality, body weight, organ weight, and gross necropsy findings. Typically, these studies are conducted to determine the lethal dose for 50% of the test animals (LD_{50}) or some other measure of mortality, and a NOAEL and LOAEL are not defined. If a NOAEL and LOAEL are defined within the range of doses tested, species differences in the dose range for testing will be manifested as species differences in NOAELs and LOAELs. For a thorough review and discussion of acute toxicity testing, see Gad and Chengelis (1998).

Acute toxicity data also can suggest the extent of absorption through different routes of exposure. If, for example, systemic toxicity or death can occur as a result of significant absorption of the chemical through dermal exposure, it must be assumed that dermal exposure also can cause reproductive or developmental toxicity.

Repeated-Dose Studies

In repeated-dose studies, animals are exposed for periods that typically range over 14, 28, 60, or 90 days. Those exposures might occur by oral, inhalation, or dermal routes. Most commonly, the studies use rats or mice, but data are sometimes available from studies in rabbits (especially for dermal exposure), dogs, or subhuman primates.

Repeated-dose studies identify the organs that are principally affected. Data also are used to define the slopes of dose-response curves and NOAELs and LOAELs for the toxic endpoints and to identify sex and species differences in toxicity at sublethal exposure. Measurements and observations in these studies include body weight, clinical signs, feed and water consumption, and parameters of clinical pathology (hematology, clinical chemistry measures of organ function). At the termination of the study, all animals undergo gross necropsy examination, and selected organs (usually liver, kidney, brain, gonads, ovaries, uterus, spleen, thymus) are weighed. Portions of all organs are preserved for histopathological examination.

Repeated-dose studies can be modified to provide valuable information on reproductive organs and, to a limited extent, on function. For example, accessory sex organs and the epididymides can be weighed in males. Sperm can be examined for concentration, motility, and abnormalities. Spermatid head counts are a useful measure of sperm production. Proper fixation, embedding, and staining of testes can permit detection of disrupted spermatogenesis (Chapin et al. 1985). Monitoring the vaginal cyclicity of females can be a useful complement to histological or endocrine data.

Changes in the weight or morphology of the reproductive organs must be interpreted in the context of other systemic or general toxicity. Such effects must be considered in the overall evaluation of reproductive toxicity, especially if there is no evidence of other systemic toxicity. The predictive value of these observations, made in 13-week rodent toxicity studies, as a screening system for reproductive toxicants has been reviewed by Morrissey et al. (1988a,b). Reproductive organ weights (testis, epididymis, cauda epididymis) and sperm motility were the most statistically powerful endpoints evaluated in males. Female cycle length was of lower predictive value because of variability within individual animals and among animals.

The dose at which toxicity to adult animals is observed in a reproductive study should be compared with the doses at which toxicity is observed in other toxicity studies. Such comparisons can be used to determine whether the pregnant or lactating female might be more sensitive to an agent than are nonpregnant or nonlactating females. The sensitivity of the paternal animal, if exposed, should also be compared.

Genetic Toxicity

Genetic damage is a possible mechanism, but not the only one, by which reproductive or developmental toxicity occurs. Although results of genetic toxicity screens alone should not be used as predictors of reproductive or developmental toxicity, genotoxicity assays such as the dominant lethal test and germ cell mutagenicity assay, or heritable translocation data, might provide supplemental information about reproductive or developmental toxicity.

Pharmacokinetics

Data on the pharmacokinetics of a particular agent, both in the species tested and in humans, can be a great aid in extrapolating toxic doses between species. Information on absorption, half-life, steady-state, and distribution with time of the parent compound and metabolites; placental metabolism and transfer; number of metabolic pathways; and comparative metabolism could be useful in predicting the risk of reproductive and developmental toxicity in humans. Such data also might assist in explaining dose-response curves for toxicity, developing more accurate comparisons of species sensitivities (Wilson et al. 1975, 1977), determining dosimetry at target sites, and comparing pharmacokinetic profiles for various dose regimens or routes of exposure. Substantial advances in the understanding of pharmacokinetics have been made, but there is still considerable uncertainty about when and how changes in human pregnancy alter pharmacokinetics (Scialli and Lione 1998). Luecke et al. (1994) and O'Flaherty et al. (1992) have developed pregnancy physiologically-based pharmacokinetics (PBPK) models that can be used to examine these changes.

Pharmacokinetic studies in reproductive and developmental toxicology are most useful if they are conducted in animals during the stages in which reproductive and developmental insults occur. The correlation of pharmacokinetic parameters and reproductive and developmental toxicity data might enhance our understanding of both the effects observed and of their predictive value (Kimmel and Young 1983; Hansen et al. 1999).

Assessments of risk for reproductive or developmental toxicity that are based on pharmacokinetic data are rarely unequivocal. One possible exception is the finding of limited or no absorption of a chemical after accidental or therapeutic exposure. For example, despite the established teratogenicity of some retiniods (e.g., Accutane) at therapeutic doses, the acceptability of all transretinoic acid (tretinoin) for dermal application was based in part on minimal or undetectable changes of the circulating concentration of endogenous retinoids after such exposure. More commonly, pharmacokinetic information will provide an improved estimate of internal dose that can be used for comparison with outcomes to provide a better basis for any estimate of risk in humans, including that for reproductive and developmental toxicity.

Because human pharmacokinetic data are often minimal, absorption data from studies of experimental animals — by any relevant route of exposure — might assist those who must apply animal toxicity data to risk assessment. Results of a dermal developmental toxicity study that shows no adverse developmental effects are potentially misleading if uptake through the skin is not documented. Such a study would be insufficient for risk assessment, especially if it were interpreted as a "negative" study (one that showed no adverse effect). In studies where developmental toxicity is detected, regardless of the route of exposure, skin absorption data can be used to establish the internal dose in the pregnant animal for risk extrapolation to human dermal exposure. For a discussion pertinent both to the development and to the application of pharmacokinetic data, risk assessors can consult the conclusions of the Workshop on the Acceptability and Interpretation of Dermal Developmental Toxicity Studies (Kimmel and Francis 1990).

Effective management of human risk is most likely accomplished through management of human exposure. Animal toxicity studies typically define a response as a function of exposure. Common descriptions of exposure (milligrams per kilogram of body weight, parts per million per hour, milligrams per cubic meter per hour, and so on) can be a poor surrogate for the toxicologically important target organ dose of the active metabolite. That is particularly true for inhalation or dermal exposure. Thus, the extrapolation of potential human risk from animal toxicity data without further knowledge of internal dosage will result in uncertainty. When it is established that a chemical or its metabolite is absorbed into the systemic circulation, the most important pharmacokinetic information will be delineation of the active agent, parent compound, or metabolite. Without this information, the usefulness of other kinetic data is decreased.

The greater the depth of understanding of toxicity and agent deposition in animals and humans, the less is the uncertainty that attends extrapolation across species and routes of exposure. When such pharmacokinetic studies are done, apparent species differences in susceptibility are often found to result from differences in absorption, fate, or elimination of the potential toxic agent rather than to biological differences in susceptibility (Renwick 1993).

Regulatory authorities typically require use of the maximum-tolerated dose (MTD) concept in nonhuman toxicological studies done in support of product registration. The use of the MTD serves to maximize response and ensure detection of all possible toxicities. However, large administered doses required to reach the MTD might result in saturation of elimination processes and exaggerated internal concentrations of the test agent. Indeed, pharmacokinetic studies showed this phenomenon for numerous therapeutic and environmental agents in the late 1960s and during the 1970s, and led to the simplistic subdivision of dose-dependent and dose-independent classifications of kinetics. Administered doses that are high enough to result in dose-independent kinetics often elicit toxic effects that are not observed at lower doses exhibiting dose-dependent kinetics (Young and Holson 1978).

It will be helpful to understand whether an agent that causes reproductive or developmental toxicity acts by exceeding a threshold concentration for a brief period of time or whether a protracted exposure is required to initiate an adverse effect. Such information could be helpful in judging the potential hazards posed by human exposure. For example, valproic-acid-induced exencephaly in the murine embryo requires that a threshold concentration be surpassed, but only for a short time. Larger total drug exposure (the area under the curve [AUC]) via infusion regimes is less active in this regard, indicating that peak concentration (Cmax) rather than total exposure (AUC) underlies the ability of this agent to induce a teratogenic response in mice. In contrast, antiepileptic therapy with valproic acid is used to maintain drug concentrations in the therapeutic range, so the Cmax in human patients is 6-10-fold less than a teratogenic concentration in mice (Nau

1986). A similar inference has been made for caffeine teratogenicity (Sullivan et al. 1987): A large single dose induced a teratogenic response, but the same amount given as four equal doses caused no malformations.

Caution in the dogmatic application of such information is necessary because a single agent can act through both pharmacokinetic exposure patterns. 2-Methoxyacetic acid, the active metabolite of 2-methoxyethanol (ethylene glycol monomethyl ether) — a chemical found in the environment — is proposed to induce murine limb malformations according to the AUC (Clarke et al. 1992). The same laboratory (Terry et al. 1994) demonstrated that, given earlier in mouse pregnancy, this agent can induce neural tube defects correlated with peak exposure. Other agents have demonstrated teratogenic activity that relies on the duration of exposure given a specific dose. For example, the teratogenicity of all-trans retinoic acid in the rat has been correlated with duration of exposure (Tzimas et al. 1997). In other cases, outcome has been correlated with both concentration and duration of exposure (Weller et al. 1999). In that study, animals exposed to short, high concentrations of ethylene oxide by inhalation on day 7 of gestation were found to have more adverse developmental effects than did animals exposed to the same concentration X time multiple but at longer, lower exposures.

Developmental toxicology studies are traditionally designed to initiate dosing at the beginning of organogenesis, which could be the wrong approach for compounds that have a long half-life because steady-state concentration is reached only after dosing for approximately four half-lives. For example, a compound with a 24-hour half-life will not reach steady-state concentration until 4 days of single, daily dose administration. In small animals with a short gestation, such as mice or rats, the embryo might reach a developmental stage of decreased sensitivity by the time steady-state concentration is achieved. Thus, pharmacokinetic information can aid in proper design of new studies or foster more accurate interpretation of results from studies already conducted.

Interpretation of data from studies in which maternal animals are exposed during lactation should account for possible interactions of the agent with maternal behavior, pup suckling behavior, and milk composition. The analysis should further consider possible direct exposure of pups via nursing, dosed feed or water, and exposure to the dam's skin, hair, or feces (EPA 1991).

In summary, the correct application of pharmacokinetic information in the judgment of risk requires a broad view of pharmacological and toxicological principles, and a thorough understanding of reproductive and developmental biology. In this regard, several general references can aid an assessment team in making the best use of the available information (Nau 1986; Nau and Scott 1987; Yacobi et al. 1993). Kimmel and Francis (1990) presented a decision-tree approach for applying pharmacokinetic information for dermal exposures that should be considered for other routes of administration as well.

Reproductive and Developmental Toxicity

Data for assessing reproductive or developmental toxicity are derived both from observations of humans and from experimental animal studies. It is beyond the scope of this document to enumerate the kinds of data that can permit a complete assessment of reproductive and developmental toxicity that covers all situations. The definition of a sufficient data set changes as scientific knowledge accumulates on specific agents and as the understanding of the predictive capabilities of animal models and other procedures improves. Appendices C and D of this document describe studies that commonly provide such information and offer guidance in their interpretation.

The reproductive and developmental toxicity data component of the evaluative process determines one or the other of two judgments: first, that the collective data are sufficient (or insufficient) to ascribe an adverse effect under specified conditions; and second, that the data are sufficient (or insufficient) to conclude that there is no adverse effect under specified conditions. To ensure systematic rigor, the process evaluates the experimental animal data and the human data independently. Each assessment uses a standard format to summarize the conditions of the test (species, dose, route, timing, duration) in which the effect (e.g., decreased sperm count, increased length of estrous cycle, altered sexual differentiation of offspring) was or was not observed.

The independent consideration of animal and human data is an initial, and incomplete, step. Only when those independent assessments are combined and integrated with analyses from the chemical and biological data section does the assessment achieve significance for health characterization purposes. Finally, the quantitative assessment of the dose-response data and integration of the toxicity and exposure information provides a characterization of risk. Chapter 3 of this report describes the integrative process.

Human Data

Given that the goal of the evaluative process is to protect human health, including reproductive and developmental processes in men and women, human data are critical.

Human studies include epidemiological investigations, clinical series, and case reports. Epidemiology is "the study of the distribution and determinants of health-related states or events in specific populations, and the application of that study to control of health problems" (Last 1995). Epidemiological studies focus primarily on human populations, although animals (as in veterinary epidemiology) or records (as in health services research) can be used as the unit of analysis. Human data can be found in a variety of sources; however, the quality of data varies from case reports to thorough epidemiological investigations.

It is important to note that not all human studies that can be used to evaluate effects meet the scientific rigor of the epidemiological method, the characteristic that differentiates an epidemiological study from others conducted in humans. The essential elements include the following:

1.   Formulation of a well-defined research question or study hypothesis that is suitable for testing.
2.   Description of the target or study population or a representative (probability) sample of it.
3.   Implementation of a standardized methodology for data collection (exposure and outcome or effect modifiers or confounders).
4.   Availability of a well-designed and descriptive analytical plan that is appropriate given the design, level of measurement, attributes of variables, and underlying statistical assumptions.
5.   Careful interpretation of the data.

A brief discussion of each feature can be found later in this chapter. Underlying the epidemiological method are two assumptions: first, that disease does not occur randomly and, second, that systematic study can identify factors that cause or can prevent disease (Hennekens and Buring 1987). Those design features help to ensure the scientific validity of human data, and they underscore the added strength and utility of epidemiological data that contribute to assessment of human health risk from toxicological hazards.

Weighing the Evidence

All epidemiological studies should be critically evaluated with respect to research design (especially in relation to study purposes), methods, analysis, and interpretation of results. Evaluation requires all aspects of the epidemiological method to be weighed carefully, as shown in Box 2-2.

Although some researchers advocate ranking studies by design type, this approach can be overly simplistic because it assumes strict adherence to methodological rigor. In essence, there is no single way to rank studies; design and methodology must be considered simultaneously. For example, a cohort study of limited statistical power should not be weighed more heavily than a well-conducted case-control study. Critical weighing of the available literature is necessary.

Selection of an appropriate control group is an important criterion for assessing case-control studies. The control groups should be similar to the study group, save for the presence of disease. Controls can be selected from registries, such as the list of people kept by departments of motor vehicles or voter registration; or they can be drawn from neighborhoods, hospitals, or lists of friends and family, depending on the study's hypothesis. Selection of appropriate controls minimizes selection bias and enhances validity of case-control studies.

For methodological aspects, greater weight should be assigned to studies that use an entire population or employ probability sampling techniques to develop a random sample. Probability samples help ensure the external validity (generalizability) of study results. There are various types of probability samples: simple, systematic, stratified, cluster, or multistage random sampling. The choice of the sample is predicated on the study's purpose.

Box 2-2   Weighing Epidemiological Evidence

Design
- Experimental
  — Randomized clinical trial
  — Community trial
- Observational (analytical)
  — Prospective cohort
  — Retrospective cohort
  — Case-control

Methods
- Population or sampling frame
- Choice of control group (for case-control)
- Choice of study exposure(s)
  — Acute vs. chronic
  — Continuous vs. intermittent
  — Dose, timing
- Choice of study outcome(s)
  — Healthy (live birth) vs. adverse
  — Unit of analysis (maternal, paternal, parental)
  — Lack of independence
- Data sources
- Standardized data collection
  — In-person interview
  — Telephone interview
  — Self-administered questionnaire
  — Existing records
- Sample size
- Participation rate

Analysis
- Multivariate analysis (Effect modification, interaction, confounding)
- Bivariate analysis
- Univariate analysis

Interpretation
- Statistical significance
- Type I and II errors
- Alternative explanations (chance, error, confounding, bias)
- Causality
  — Necessary, sufficient
  — Risk factors

Epidemiological studies that collect data from several sources are useful because their validity and reliability can be further assessed. Such studies should be given greater weight than those that rely exclusively on self-reported data. The use of biological markers of exposure, susceptibility, or effect in disease is another strategy for maximizing the validity of study results. However, careful interpretation of biomarker data is needed, given our limited understanding of what the findings might actually mean in terms of human health.

Selection of a health outcome for study depends in part upon the exposure of interest and methodological considerations such as the ability to define, measure, and validate adverse outcomes, especially if self-reported. Operational definitions for outcomes can be general or specific in nature (e.g., all birth defects versus spina bifida, respectively), and will affect the type of statistical analysis which can be performed and the interpretation of results. Statistical power may be limited if restrictive operational definitions are used for rare outcomes, or if other important covariates cannot be fully addressed.

Use of a standardized methodology for ascertaining data on exposure, outcome, or other relevant covariates is an essential feature of an epidemiological study that enhances the validity of results. All study participants should be subjected to the same method for collecting data. In-person interviews are reported to provide the most reliable self-reported exposure data, followed by telephone or mail survey techniques. Standardized forms for collecting existing data also should be used.

Epidemiological studies of sufficient size to minimize Type I (alpha) and II (beta) errors should be weighed more heavily than statistically underpowered studies. Type I error is the incorrect rejection of the null hypotheses — the investigator erroneously concludes that an association exists. Alpha levels, by convention, are typically set at .05, which denotes that a "significant" chance finding can occur 5% of the time. Type II errors occur when the investigator fails to reject the null hypothesis when an association does exist. Beta levels, by convention, are typically set at .20, yielding a study power of 80% — a study detects a true difference 80% of the time. Ad hoc power calculations can provide better insight about the sufficiency of the sample for study purposes. However, many published papers are secondary analyses with uncertain statistical power. As such, the absence of an effect needs to be weighed in relation to the study's statistical power and interpreted accordingly.

The analytical plan of epidemiological studies should use descriptive and analytical techniques in describing the sample and results. Descriptive statistics, such as frequency distributions, cross-tabulations, measures of central tendency, and variation, can help explain underlying distributions of variables and direct the assessment of appropriateness of more advanced statistical techniques. Careful weighing of study findings with respect to the design and methods helps to ensure the validity of results.

Greater weight should be assigned to epidemiological studies that have carefully assessed statistical significance. Because a wide variety of statistical tools are available for testing significance, consideration must to be given to the design of the study, to the types of data collected, to the sample size, and to the study purpose. Several textbooks provide diagrams to assist in selecting the appropriate statistics (e.g., Hennekens et al. 1987). Studies that provide confidence intervals and not just probability values alone should be assigned more weight.

The chance of Type I and II errors should be considered. Alternative explanations for the results should be carefully addressed. Studies that discuss results in relationship to chance findings, random errors, possible confounders, or sources of bias should be weighed more heavily than are studies that ignore or incompletely address those issues.

Case Reports and Clinical Series

Almost all exposures that are currently recognized as having unequivocal developmental or reproductive toxicity in humans were initially recognized in case reports and clinical series. This was possible because adverse reproductive exposures typically produce qualitatively distinct patterns of toxicity. They do not normally affect all reproductive and developmental outcomes indiscriminately. This effect is most apparent with developmental toxicity — exposure to a particular agent during development characteristically causes a distinctive pattern of congenital anomalies depending on the timing of exposure.

Clinical series can be compelling when they demonstrate the occurrence of a highly characteristic pattern of anomalies in children of women who experienced similar well-defined exposures at similar times in pregnancy. The association is especially convincing if both the pattern of anomalies and the exposure are rare in other circumstances. For example, the characteristic patterns of congenital anomalies produced by excessive maternal exposure to alcohol, toluene, methylmercury, or polychlorinated biphenyls during pregnancy led to recognition of the developmental toxicity of exposure to these substances. The dysmorphic syndromes that occur—fetal alcohol syndrome, toluene embryopathy, congenital Minimata disease, and congenital rice oil disease, respectively—are not distinguished by the presence of a single distinctive feature. In fact, many of the component features are rather common and can have a variety of causes. When the features occur together, however, they constitute a distinctive pattern of congenital anomalies that is rare except in children born to mothers who have been exposed to one of these substances during pregnancy.

In contrast to well-designed cohort and case-control studies, neither case reports nor clinical series can provide reliable quantitative estimates of the risk of adverse outcome in children of women with a toxic exposure during pregnancy. Case reports and clinical series are useful as a means for generating hypotheses that can be tested with analytical designs. Adverse reproductive outcomes are common in the general population—spontaneous abortion occurs in 15-20% of recognized pregnancies, and approximately 5% of all children have serious congenital anomalies or mental retardation that become apparent within the first year of life. The frequency of learning disabilities and behavioral disorders in childhood is even greater. Coincidental occurrence of various exposures in a pregnant woman and miscarriage or congenital anomaly in the offspring is, therefore, common. Chance associations are even more likely if one considers the full range of possible adverse reproductive effects and exposure of either parent for a variable period of time before conception. The observation of adverse developmental or reproductive outcomes in a few case reports or clinical series is, therefore, never sufficient by itself to establish the reproductive toxicity of an exposure in humans.

Assessing Causality in Human Studies

Careful attention must be given to assessing causality. Causation also can be ranked in terms of weight of evidence (Jekel et al. 1996). Greater weight is given to a sufficient cause that, when present, always results in disease. Next, a necessary cause also precedes disease and, when absent, cannot result in disease. The third and weakest type of causation is a risk factor that when present, increases the likelihood of an outcome in exposed versus unexposed individuals. A risk factor is neither a necessary nor a sufficient cause of death or adverse health outcome. Most observational studies estimate risk factors in assessing causality. Necessary and sufficient causes are often useful for the study of infectious diseases.

If an association is observed between exposure to a specific substance and a particular adverse outcome, the investigator must determine whether it is a chance finding or causal in nature. There are several widely recognized criteria for assessing causality, including temporal relationships, strength of association, dose-response relationships, replication of findings, biological plausibility, consideration of alternative explanations, cessation of exposure, and specificity of association (Gordis 1996). Experimental and observational designs alike can assess temporal relationships; descriptive studies cannot.

The strength of an association and dose-response relationships are determined in an observational epidemiological study by assessment of the relative risk (RR) for cohort studies or odds ratios (ORs) for case-control studies. RRs or ORs greater than 1.0 denote an increase in risk of disease given exposure in comparison with the unexposed. Conversely, a RR or OR of less than 1.0 denotes a reduction in risk of disease given exposure. RRs or ORs equal to 1.0 denote no effect. Confidence intervals that exclude 1.0 indicate statistical significance of the risk factor. Multivariate modeling or stratification procedures can be used in observational studies to assess for confounding, interaction, or effect modification and, thereby, to help rule out alternative explanations.

The existing literature is used to assess the remaining criteria for causality: replication of findings in other populations, biological plausibility to aid in interpreting risk factors, cessation of exposure, specificity of association, and consistency with other knowledge. In attempting to assess causality based on the available data, careful consideration must be given to the direction and strength of statistical estimates across studies that use various populations and study designs. Formalized approaches currently are available to weigh empirical findings (e.g., meta-analysis) and should be considered. The actual interpretation of the empirical evidence must be tempered by study limitations and attention to methodological rigor. If human data are considered with animal data, the plausibility of the agent causing an adverse outcome could be enhanced. See Appendix C for further discussion of causality.

Evaluating human studies requires careful assessment of all elements of the epidemiological method. This is not an easy process, and it requires understanding of research methodology and appreciation of biostatistics.

Experimental Animal Toxicity

Utility and Limitations

The study of chemical exposure in experimental animals is a reasonably efficient and effective means for ascertaining a chemical's toxic potential in humans. Investigators have developed a standard series of animal test procedures that domestic and international regulatory bodies require, for example, for approval to market drugs, pesticides, and, to a lesser degree, other industrial and commercial substances. Data showing adverse effects from such animal reproductive and developmental toxicity studies are assumed to be predictive of a potential human reproductive or developmental effect, although the precise manifestations may not be the same. This assumption is based on comparisons of data from animals and humans for exposures that are known to cause human reproductive and developmental toxicity (Thomas 1981; Nisbet and Karch 1983; Kimmel et al. 1984; Hemminki and Vineis 1985; Meistrich 1986; Working 1988; Francis et al. 1990; Newman et al. 1993). Animal models are available for all known exposures causing reproductive toxicity in humans, and in many cases the effects in animals are similar, or are of a similar type, as those observed in humans (Kimmel et al. 1984,1990; Schardein 1998). The species that show strict concordance with human effects vary (Schardein 1998), but this is likely due to variability in human exposure, the level of exposure, and differences in pharmacokinetics and pharmacodynamics that are to some extent chemical-specific.

Despite the proven utility of animal and other laboratory data, several factors can limit their usefulness. There are close similarities among mammals in such biological processes as fetal and embryonic development, sperm production, and ovulation, but distinct differences and variations also exist among species. Such differences can limit the certainty of predicting that an effect seen in a laboratory species will occur in humans. It is not uncommon, for example, for one animal species to exhibit an adverse effect, while a second species either shows no effect or shows effects only at markedly different doses. There are practical restraints on the number of animals that can be studied; this places statistical limits on the certainty of some test results. Poor study design or laboratory practices also can compromise the data. Thus, because experimental animal toxicity data can be misinterpreted, it is imperative that the evaluative process include a review and interpretation of animal data by scientists with appropriate training and experience. The logic that underpins their interpretation of data should be stated clearly in the evaluation so that other experts can understand the basis for the evaluative judgment.

Adverse Effect

In general, three criteria must be met to support a conclusion that animal data are sufficient to indicate an adverse effect in the species studied under the conditions specified for the experiment:

- *At least one well-conducted study must show reproductive or developmental toxicity in a mammalian species.* When the study data are insufficient, improper study design or execution, inadequate doses or duration of exposure, poor survival, or too few animals to achieve statistical power are often the cause. At present, no nonmammalian or in vitro systems are considered to be predictive of human responses, and are not accepted by regulatory agencies for human hazard assessment of reproductive and developmental toxicity.

Studies might be considered adequately conducted but still insufficient because the endpoints are not clearly related to an adverse effect. Such data should still be cited. For example, only one study might be relevant to reproductive toxicity. That study might have noted a decrease in production of progesterone by cultured granulosa cells. Although the study is adequate in every technical respect, the data themselves are insufficient for rendering an assessment of animal hazard because the relationship of this effect to an adverse effect in vivo cannot be predicted. In such an instance, the evaluators should consider more definitive test data as a critical data requirement for predicting effects on ovarian function and related outcomes in humans.

- *The data must be interpreted as having biological significance.* Although the evaluative process strongly endorses the application of appropriate and rigorous statistical methods, it must be clear that, when a study meets conventional statistical criteria, it also must yield data that reflect an effect that is both biologically plausible and considered adverse.

In the occasional instance where there is statistical but not biological significance, the evaluation must clearly articulate the basis for concluding that the evidence is insufficient to show an adverse effect, and discuss the uncertainties associated with the data. A case in point is the evaluation of HFC-134a exposure in a two-generation study by Alexander et al. (1996) in which the parental generation was exposed before mating and during pregnancy and lactation, and F2 offspring were found to have slight but statistically significant delays in physical and reflex development that were not clearly dose-related. Because the F2 generation was never exposed to HFC-134a directly or indirectly and the changes reported represented a one-half to one-day delay and were not clearly dose-related, these changes were not considered treatment-related (see Appendix A for further discussion).

- *Dose response.* Evidence of a dose-response relationship is an important criterion in the assessment of a toxic exposure in experimental animal studies. However, traditional dose-response relationships might not always be observed for some endpoints. With increasing dose, for example, a pregnancy might end in a fetal loss rather than in birth of live offspring with malformations.

Typically, the demonstration of *no adverse effect* requires a larger set of evidence than does the demonstration of an adverse effect. To support a conclusion that a given exposure does not cause developmental toxicity, the available studies must be conducted in at least two mammalian species and must test for a wide variety of pre- and postnatal outcomes (EPA 1991). A minimum data set for a conclusion of no reproductive toxicity would normally consist of at least one two-generation reproductive toxicity study.

Additional studies often are warranted, especially when there is prior knowledge of the general toxicity of a given agent or chemical class, or when there is knowledge of the pharmacological activity of the agent. Several examples illustrate cases in which the minimum data set described above should not be used as the basis for concluding no adverse effect.

- The absence of a postnatal functional evaluation in the database renders the developmental toxicity database incomplete without an additional developmental neurotoxicity study (Moore et al. 1995a; EPA 1998c).

- A standard reproductive study in rats showing no effect on the ability of males to impregnate females should not be considered to support a conclusion that there is no male reproductive toxicity in all species. Unlike humans, rodents produce sperm in numbers that greatly exceed the minimum requirements for fertility. A substantial reduction in sperm production in rodents may not compromise fertility in rodents while a less severe reduction in human males could cause reduced fertility.

- If there are data that suggest that standard experimental animal species are not appropriate for comparison with humans, data from a less commonly used species, such as dog or a nonhuman primate that is metabolically similar to humans, would be needed to confirm the lack of reproductive and developmental toxicity.

- As they are performed today, in vitro studies will not by themselves provide sufficient evidence of no adverse effect.

- Studies in two species are often available in which pregnant females were exposed during pregnancy and killed just before parturition. This permits full evaluation of adverse effects on mother and fetus. Such prenatal developmental toxicity studies designed to determine a substance's potential to cause structural abnormalities, growth deficits, or death are available in two species for many agents, drugs and pesticides, in particular. When such studies demonstrate no adverse effects, one might conclude that the data are sufficient to indicate that there is little or no risk that the agent might cause developmental toxicity manifested at birth. However, it is important to recognize the limitations of these studies and that if dosing stops at the end of major organogenesis (gestation day 15 in rats and mice, day 19-20 in rabbits), later fetal exposures might result in further growth retardation and in other developmental defects (e.g., those occurring in late-developing reproductive organs such as hypospadias). In addition, there is no information from such studies on the postnatal effects of prenatal exposures, including possible neurobehavioral deficits or other impaired organ system function.

Similarly, the absence of adverse effects in a two-generation reproductive study would not preclude the possibility of significant reproductive toxicity that is not manifested as a fertility problem, unless more detailed sperm evaluations, estrous cycling, ovarian histology, and endocrine function were included. Detailed description of animal testing protocols, and their qualifications and limitations are discussed in Appendix D.

3

The Evaluative Process:
Part II. Integration of Toxicity and Exposure Information

The integration step of the evaluative process is conducted in three stages. In the first stage, the evaluators examine the data for relevance to potential human toxicity. Then, if the data are determined to be relevant to human exposures, a quantitative assessment is conducted. Finally, the concluding step of the evaluative process is the integration of toxicity and exposure information to characterize the risk of potential reproductive and developmental toxicity. A narrative summary communicates to Navy environmental health practitioners the scientific judgment on a chemical's risk for reproductive and developmental toxicity.

INTERPRETATION OF TOXICITY DATA

The interpretive section of the evaluative process considers all relevant information in the course of reaching a judgment about whether an exposure has the potential to cause developmental or reproductive toxicity in humans. In most cases, animal data are considered relevant indicators of human risk, unless there is modifying information that suggests otherwise. The most common reason for concluding that no hazard exists for humans is the availability of sufficient experimental data that reveal no adverse effects in animal studies. Some experimental data might demonstrate toxicity of limited relevance to humans because of species differences in metabolism or sensitivity, lack of probable human exposure, or human evidence of no effect. Animal data in which no adverse effects are observed do not always preclude human effects, nor do adverse effects in animals inevitably predict human toxicity.

The evaluative process requires an integrated consideration of a variety of data. Integration involves combining the summary statements formulated during the review of animal and human reproductive and developmental toxicity data and considering them in the context of systemic toxicity parameters and pharmacokinetic data. A weight-of-evidence approach is then used to formulate judgments about the potential for human hazard. In this process, the evaluators develop separate statements to address developmental toxicity, female reproductive toxicity, and male reproductive toxicity. In each case, the basis for the judgment is articulated and makes particular note of such critical factors as replication of effect across species, exposure routes, dose-response parameters, relationship of effective dose to doses that cause other forms of toxicity, and comparative metabolic data.

To achieve a degree of consistency in the interpretation of experimental animal data, this report uses "relevance" terms:

- *Irrelevant* means that pharmacokinetic or mechanistic features of the experimental animal model are known in detail and are demonstrably inconsistent with human exposure or response.
- *Relevant* identifies a data set in an experimental animal species for which pharmacokinetic and mechanism information is adequate to demonstrate a particular similarity to humans.
- *Assumed relevant* indicates there is no modifying supplemental information.

For many agents, there is no detailed understanding of absorption, distribution, biotransformation, or excretion in experimental animals or humans. In these cases, studies of the most sensitive experimental animal species are assumed to be relevant, and would thus drive the judgment of potential risk to humans.

Where possible, an evaluation should use pharmacokinetic data, including metabolic and mechanism-of-action information, to determine the relevance of experimental data to humans. Should the available data for a particular species demonstrate a pharmacokinetic pattern similar to that found in humans, the data from that species will be considered *relevant*. But if, for example, an agent given to an experimental animal requires biotransformation to produce toxicity, and if humans are known to be incapable of that biotransformation pathway, then toxicity data from that experimental animal species would be considered *irrelevant* to humans.

Toxicity always depends on exposure conditions, such as route of administration, timing and duration of administration, and dose. The conservative default assumption is that, without data to the contrary, treatment of an experimental animal by any route is *assumed relevant* to human exposure by any route. That default assumption can be dropped when adequate modifying information is available. If, for example, an experimental animal study uses oral dosing, and humans are known not to absorb the agent by the oral route, then the experimental data are *irrelevant* for human oral exposure (but not necessarily for other routes of human exposure). In another example, if experimental animals are exposed to an agent via the oral route and adverse effects are observed, but exposure to humans is topical and the agent is not absorbed through the skin, then the experimental data are *irrelevant* to humans.

The use of a template (Box 3-1) for summarizing the available data is advised as a guide for ensuring consistency in the characterization of reproductive and developmental hazards.

Default Assumptions To Be Considered in Assessing Reproductive and Developmental Toxicity Risk

Certainty of judgment about toxicity is in large part proportional to the quality and amount of chemical-specific data. In many instances, the desired data are not available; in such circumstances, it has been traditional to adopt default assumptions and proceed with the assessment. Default assumptions should incorporate all available information to reduce the level of uncertainty as much as possible. It

Box 3-1 Template for Summarizing Reproductive and Developmental Toxicity Using the Evaluative Process

There is _____ [sufficient, insufficient] _____ evidence in

[humans and/or animals] _____ that _____ [chemical/agent] _____

_____ [does or does not] _____ cause _____ [reproductive toxicity in

males/females, developmental toxicity] _____ when exposure is

_____ [route, dose range, timing, duration] _____. Relationship to adult

toxicity stated. The data are _[relevant, assumed relevant, irrelevant]_

to consideration of human risk.

might be necessary to choose from a range of reasonably plausible default values, such as the volume of inhaled air for the sedentary individual, the worker who performs physically demanding tasks, or the active jogger or marathon runner. In such instances, the common practice is to choose assumptions that would estimate the upper range of exposure in individuals who constitute the exposed population. In cases in which there is little or no information available, the assumption selected might deliberately represent a worst-case value. In every case, default assumptions should be used sparingly and openly, with full disclosure of the degree of certainty. The general default assumptions proposed for use in this evaluative process are summarized below.

Absorption

Rates of absorption and elimination are assumed by default to be comparable among species. If experimental animal absorption has been determined but human absorption is unknown, human absorption is assumed to be the same as that in the species with the highest degree of absorption.

When quantitative absorption data for a route of exposure indicate differences between humans and the relevant test species, the no-observed-adverse-effect level (NOAEL) might need to be adjusted proportionately.

Cross-Species Extrapolation

When assessing manifestations of toxicity, evaluators might base their conclusions about relevance on the mechanism that produces a toxicological effect; however, a basic default assumption is that any manifestation of reproductive or developmental toxicity is relevant to humans unless the mechanism by which it occurs is impossible in humans. For example, if a toxic effect occurs in animals through an inhibition of folic acid synthesis, that effect would not be considered relevant for humans because humans do not synthesize folic acid. It is unusual, however, to have such detailed knowledge about mechanisms of toxicity from experimental animal studies.

It should be noted that the particular effect produced in an experimental animal study does not generally have a bearing on determinations of relevance. If an agent causes tail defects in the offspring of treated mice, for example, this effect is not automatically considered irrelevant to humans simply because humans do not have tails. Instead, the assumption is that the mouse study demonstrated that the chemical substance interfered with vertebral development and therefore has relevance for vertebral or other features of human development.

Where there are experimental data from more than one species, the default assumption is that humans are at least as sensitive as the most sensitive animal species. If the data indicate, however, that some particular species is a more relevant surrogate for humans, either because of physiological similarity at the site of action or because of the pharmacokinetic parameters associated with the substance under review, such information will preempt that general assumption.

In the absence of data, activation and detoxification pathways in animals and humans are assumed by default to be qualitatively and quantitatively similar.

Adjustments of NOAELs from inhalation exposure studies to a human equivalent concentration, based on adjustments for minute volume, respiratory rate, and other factors, are appropriate for reproductive and developmental toxicity (EPA 1994). Toxicity data are scaled directly from experimental animals to humans on the basis of minute volume per kilogram of body weight for inhaled materials and by weight (or volume) of the dose per body weight or surface area for other routes of exposure. The first priority is to use the internal dose at the target site, if available.

Exposure Duration

Evaluators must assume that a single exposure at a critical time in development or in the reproductive cycle might produce an adverse effect; that is, repeated exposure is not necessary for reproductive or developmental toxicity. The fact that toxicity might be cumulative with repeated exposure is another important consideration. In most cases, the data available for reproductive and developmental toxicity risk assessment are from studies that use repeated exposures. The NOAELs and lowest-observed-adverse-effect levels (LOAELs) for reproductive and developmental effects are, however, usually based on a daily dose (e.g., milligrams per kilogram body weight per day (mg/kg/d)). When extrapolating duration data from animals to humans, it is important to consider not only the percentage of the life span during which exposure occurs, but also physiological (including developmental) time of the life span. Rodent development is much more rapid than in humans and maturity at birth differs considerably in rodent species and humans. For example, the duration of exposure in a prenatal developmental toxicity study that extends from gestation day 6 to gestation day 15 in rats might be expressed as equivalent to exposure from approximately 1 week to 8 weeks gestation in humans. Developmental stage also varies among organ systems for experimental animals and humans, both pre- and postnatally; therefore, relative developmental stage must be considered for the organ system(s) of concern. For example, nervous system development at birth in humans is approximately equivalent to that 5 days postnatally in rat pups, while respiratory and cardiovascular function are more similar at birth due to the demands of the ex utero environment.

Windows of Vulnerability

The concept of windows of vulnerability for developmental toxicity is generally known and accepted (Wilson 1973; Moore 1988). For example, thalidomide causes abnormal ear development, autism, duplication of thumbs, and heart and renal defects after exposure on gestation days 20-24 in humans; shortened or missing limbs after exposure on gestation days 24-33; and rectal stenosis and triphalangism of the thumbs on gestation days 35-46 (Schardein 2000; Strömland et al. 1994; Miller and Strömland 1999). Screening studies in animal models usually involve treatment throughout organogenesis and, in some cases, throughout fetal and early postnatal life, so that critical periods of vulnerability can not always be discerned. When effects are seen, additional follow-up studies may be conducted to more clearly delineate the time period during development when the effects are produced. For example, exposure of humans to angiotensin converting enzyme (ACE) inhibitors results in oligohydramnios, renal pathology, fetal growth restriction, hypoplastic ossification of the skull, and, possibly, patent ductus arteriosus — unexpected findings based on the standard animal testing studies. When follow-up studies were conducted, it was shown that these effects resulted from exposure during the second and third trimester in humans, equivalent to the late fetal and early neonatal stages in rodents (Barr 1997). Screening studies conducted in animals using repeated exposures throughout organogenesis may also be followed up with more discrete exposures to determine critical windows. For example, the finding that boric acid caused a number of skeletal alterations in studies in rats, mice, and rabbits with exposure throughout organogenesis (Heindel et al. 1994; Price et al. 1996), was followed up with studies by Narotsky et al. (1998) that show the specific times for the induction of skeletal alterations at different sites in the axial skeleton. Information on critical windows of exposure for several organ systems has recently been summarized (Selevan et al. 2000).

Additivity

Exposure by multiple routes is assumed to be additive. The default assumption is that simultaneous exposure to multiple agents having the same site or mode of action results in additive effects. Thus, for example, estimates of the developmental toxicity of chlorinated dibenzodioxins and dibenzofurans should consider the use of toxic equivalency factors (Safe 1993), provided the quantitative value assigned to each congener is relevant to the toxic effect under consideration.

QUANTITATIVE EVALUATION

Once an assessment has determined that the data indicate human risk potential, the next step is to perform a quantitative evaluation. Here, dose-response data from human and animal reproductive and developmental toxicity studies are analyzed to select LOAELs and NOAELs or to calculate a BMD. The assessment should use quantitative human dose-response data if the data span a sufficient range of exposure. Because data on human dose-response relationships are rarely available, the dose-response evaluation is usually based on an assessment of data from tests performed in experimental animals.

Box 3-2 defines terms commonly used in quantitative evaluations.

Identification of the NOAEL and LOAEL

The dose-response evaluation defines the range of doses that produce reproductive and developmental toxicity, the routes of exposure, the timing and duration of exposure, the species specificity of effects, and any pharmacokinetic or other considerations that might influence comparison with human exposure. Much of the focus is on identification of the adverse effect observed at the LOAEL and the NOAEL for the study.

Box 3-2. Definitions

No-observed-adverse-effect level (NOAEL). The NOAEL is the highest dose at which there is no biologically significant increase in the frequency of an adverse reproductive or developmental effect when compared with an appropriate control group. Biological significance is based on expert judgment and consideration of statistical analyses.

Lowest-observed-adverse-effect level (LOAEL). The LOAEL is the lowest dose at which there is a biologically significant increase in the frequency of adverse developmental or reproductive effects when compared with the appropriate control group. Biological significance is based on expert judgment and consideration of statistical analyses.

Uncertainty factors (UF). A UF is a value applied to a NOAEL to account for variability in response across species and among humans. It usually is a factor of 10 for each area of variability (uncertainty), although each factor might be reduced or enlarged according to the quality and amount of data. Additional factors may be applied to account for uncertainty due to missing or inadequate data. A factor of 10 is also commonly applied when the data identify only a LOAEL instead of a NOAEL.

Unlikely-effect level (UEL). This is an estimate of the daily exposure of a human population that is assumed to be without appreciable risk of causing reproductive or developmental effects. The duration can vary, depending on the human exposure scenario of interest.

Margin of exposure (MOE). The MOE expresses the magnitude of difference between a level of anticipated human exposure and the highest level at which there is no significant increase in the frequency of an adverse effect (NOAEL). The MOE is the ratio of the NOAEL for a specific toxic effect to the estimated human exposure.

Benchmark dose (BMD). The BMD is used as an alternative to the NOAEL for reference dose calculations. The dose response is modeled and the lower confidence bound for a dose at a specified response level is calculated. For a further description, see the section on BMD calculation.

Human equivalent concentration (HEC). The HEC is used to describe the dose of an agent to which humans are exposed through inhalation. The HEC is the estimated concentration that is equivalent to that used in an experimental animal species. The HEC is estimated using adjustment factors that account for such species-dosimetric differences as ventilatory parameters and lung surface areas, as well as factors related to the gas, aerosol, or particulate nature of the agent.

Determination of the NOAEL is based in part on the statistical evaluation of the data from relevant studies. The statistical procedures used should reflect both the design of the experiment and structure of the data (for example, a blocked experiment in which the observed individuals are fetuses, but exposed individuals are their dams), and the type of data (dichotomous, categorical, or continuous). An overall evaluation of the statistical significance of treatment effect should be carried out first, ideally using a trend test such as Williams's Test (1971, 1972) or the NOSTASOT (no statistical significance of trend) procedure (Tukey et al. 1985), because such tests tend to be more powerful than analysis of variance (ANOVA)-like tests when the alternative to the null hypothesis (of no treatment effect) is a monotonically increasing or decreasing response. In the presence of a significant overall treatment effect, a NOAEL is determined to be the highest dose level at which there are no significantly different responses from the control group. This can be determined using sequential trend tests, sequentially dropping the highest dose in the remaining set of doses until there is no significant trend, as in the NOSTASOT procedure (Tukey et al. 1985; Faustman et al. 1994), or Williams's test (Williams 1971, 1972), or by conducting pairwise tests that appropriately control the overall Type I error, such as Dunnett's test for independent, continuous data (Dunnett 1955, 1964). It should be recognized that the power of reproductive and developmental toxicity studies to detect effects with typically 20 animals per dose group varies for endpoints within and between studies. For example, the likelihood of detecting significant changes in continuous endpoints such as fetal weight is much greater than the likelihood of detecting significant changes in malformations (Nelson and Holson 1978). It is beyond the scope of this document to discuss statistical methods for evaluation of data in detail. Study types can be of varied design, thereby influencing statistical approaches for evaluation. Statistical methods for a variety of toxicology study designs are available (Gad 1998).

The NOAEL and LOAEL are constrained by the exposure concentrations used in a given experiment. For example, consider the case in which the administered doses used in an experiment were 1, 10, 100, and 1,000 mg/kg/d and the highest dose at which no increase in adverse effect (the NOAEL) was seen was 10/mg/kg/d. The exposure concentration at which no toxicity would have occurred might have been at any dose between 10 mg/kg/d and just below 100 mg/kg/d, but the NOAEL would still be 10 mg/kg/d, because no other doses between 10 and 100 mg/kg/d were tested. The NOAEL approach also neglects the shape of the dose-response curve and ignores information about responses that were obtained at higher doses (Figure 3-1), because only the dose at the NOAEL is used. Finally, the NOAEL is dependent on the statistical power of the study, so that using larger numbers of animals might have allowed for an effect to be detected at 10 mg/kg/d, leading to a lower NOAEL.

Calculation of the Benchmark Dose

Because the literature describes several limitations in the use of NOAELs (Gaylor 1983; Crump 1984; Kimmel and Gaylor 1988), the evaluative process considers other methods for expressing quantitative dose-response evaluations. In particular, the BMD approach originally proposed by Crump (1984) is used to model data in the observed range. That approach was recently endorsed for use in quantitative risk assessment for developmental toxicity and other noncancer health effects (Barnes et al. 1995). The BMD can be useful for interpreting dose-response relationships because it accounts for all the data and, unlike the determination of the NOAEL or LOAEL, is not limited to the doses used in the experiment. The BMD approach is especially helpful when a NOAEL is not available because it makes the use of a default uncertainty factor for LOAEL to NOAEL extrapolation unnecessary.

The BMD is a model-derived estimate of a particular level of response above background for dichotomous endpoints that is near the lower limit of the range of experimentally detectable effects, the benchmark response (BMR; e.g., 5% or 10%). To obtain the BMD, one begins by modeling the data in the observed range, resulting in a curve that, for dichotomous data, gives the probability of response for the experimental dose corresponding to the BMR. The BMDL is the lower confidence limit on that dose and is the value used for calculating reference levels.[1] Figure 3-2 illustrates the relationship between the dose-re-

---

[1]The convention of using BMDL as the lower confidence limit follows the terminology proposed in the paper by Crump (1995); this has also been adopted for use in the EPA BMDS software, since it refers explicitly to the lower confidence limit value.

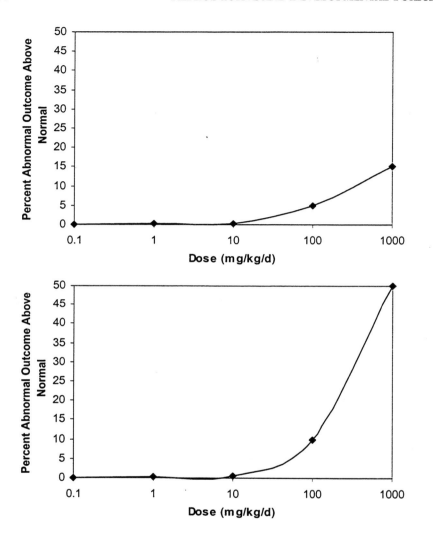

FIGURE 3-1 Sample dose-response curves. In each case, the NOAEL is 10 mg/kg/d and the LOAEL is 100 mg/kg/d, assuming that the increase above control at these exposures is significant.

sponse model, the BMD and the BMDL for a BMR of 10% above background for a dichotomous endpoint.

Using the BMD approach, one can calculate a value for each effect of an agent for which sufficient data are available. A level between the

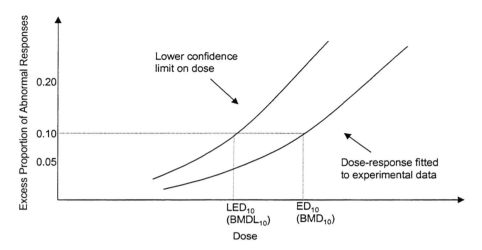

FIGURE 3-2 Illustration of the BMD. LED_{10} = $BMDL_{10}$, lower confidence limit on the dose resulting in a 10% response; ED_{10} = BMD_{10}, best estimate of the dose at a 10% level of response derived from the model. Source: Adapted from Moore et al. (1995a).

BMD_{01} and the BMD_{10} usually corresponds to the lowest level of observed risk that can be estimated for dichotomous endpoints without extrapolating to lower levels. The articles by Allen et al. (1994a, b) and an EPA background document (1995) provide a broader discussion of these issues. The U.S. Environmental Protection Agency (EPA) has free software for modeling the BMD and BMDL (http://www.epa.gov/ncea/bmds.htm).

For continuous data, there are several options for deriving the BMR: (1) the degree of change considered adverse for that effect is used as the BMR and the data are modeled as continuous data; (2) if individual data are available and there is an accepted level of change considered adverse, the data can be "dichotomized" (number above or below the cutoff value, perhaps based on some quantile of the distribution), and modeled as for dichotomous data; or (3) in the absence of any knowledge of what to consider adverse, a standard can be applied (e.g., one standard deviation of the control mean), and data handled as in 1 or 2 above for the degree of change. An alternative to modeling continuous data directly is to use the so-called "hybrid" approach, such as that described by Gaylor and Slikker (1990), Kodell et al. (1995), and Crump (1995). This approach fits continuous data using continuous models, then presuming a distribution of the data, a BMD and BMDL can be calculated in terms of the fraction affected, resulting in a probability (risk) of an individual being affected as for dichotomous responses.

Duration Adjustment

Adjustments are often made in the NOAEL or BMD to account for the exposure scenario of concern. In the case of inhalation exposure, for example, if a study involved exposure to 500 parts per million (ppm) for 6 hours per day (h/d), and there are no modifying pharmacokinetic data, the adjusted NOAEL or BMD for a continuous exposure would be calculated by multiplying by 6/24, yielding 125 ppm. Adjustment to account for the duration of exposure has not been applied routinely in assessments of developmental toxicity; such an adjustment is made in the case of assessments for reproductive toxicity. The Subcommittee on Reproductive and Developmental Toxicology recommends that exposure duration should be considered in developmental and reproductive toxicity assessments alike. The reason for this recommendation is that adjusting for duration of exposure is likely to be more conservative with repeated exposures than with single exposures, even for developmental toxicity data (Weller et al. 1999). In the case of occupational exposure during a 6-8 h workday, this adjustment could be unnecessary. However, if pharmacokinetic data indicate accumulation with repeated exposure, an adjustment would be appropriate.

Pharmacokinetic information that relates blood concentration to toxic response is critical in defining such dose-response relationships, but information on peak blood concentrations or blood concentrations over time (area under the curve (AUC)) is seldom available. One agent for which such information has been published is 2-methoxyacetic acid (2-MAA), the active metabolite of 2-methoxyethanol. Terry et al. (1994) showed that peak concentration was related to neural tube defects observed after exposure in mice on gestation day 8, whereas area under the curve was shown to be related to limb defects after exposure to 2-MAA on gestation day 11 (Clarke et al. 1992), suggesting that the time of exposure and pattern of development of the susceptible organ may be as important as the dose metric. Kimmel and Young (1983) showed that a combination of peak exposure and area under the curve for a single dose of salicylic acid was important in defining the dose-response relationship for malformations and other effects. Less has been done to examine the relationship with longer-term, repeated dosing. Pharmacokinetic data may be used to adjust exposure concentrations for such agents. As more pharmacokinetic information becomes available, it is important to minimize the dependence on default assumptions and to encourage the use of pharmacokinetic data to determine the appropriate dosimeters to use in adjusting exposure levels and determining internal dose.

Uncertainty Factors

Factors to account for various uncertainties are applied to the NOAEL, LOAEL, or BMD to derive a UEL. The total size of the uncertainty factor (UF) varies, accounting for assumed or known interspecies differences, variability within humans, quality and quantity of the data, consistency, slope of the dose-response curve, background incidence of the effects, and pharmacokinetic data. The relevance of the species, type of effect, dose, route, timing, and duration of exposure are additional factors that might influence its size. A discussion of UFs is provided in several papers (e.g., Lewis et al. 1990; Renwick 1991, 1998; Dourson et al. 1996; Renwick and Lazarus 1998).

UFs for reproductive and developmental toxicity applied to the NOAEL often include 10-fold factors for interspecies and intraspecies variation. Additional factors might be applied to account for other uncertainties or for additional information that might exist in a database. For example, in circumstances in which only a LOAEL is available, it might be necessary to use an additional UF uncertainty factor of up to 10, depending on the sensitivity of the endpoints evaluated, the adequacy of the tested dose, or general confidence in the LOAEL. An additional uncertainty factor of 3-10 has been used by EPA (1996a) to account for database deficiencies, particularly the lack of reproductive and developmental toxicity studies.

The experience gained from assessing the data-rich chemicals lithium and boric acid using the evaluative process described by J.A.

Moore et al. (1995b, 1997) showed that the expert review groups did not routinely apply factors of 10 for interspecies and 10 for intraspecies variability to the NOAEL. In each case, interspecies and intraspecies factors were reduced by half a log (J.A. Moore et al. 1995b, 1997); knowledge of pharmacokinetics was useful in reducing uncertainty in the data for predicting human risk.

Calculation of the Unlikely Effect Level

The UEL for reproductive and developmental toxicity is derived by applying uncertainty factors to the NOAEL, LOAEL, or BMDL. To calculate the UEL, the selected UF is divided into the NOAEL, LOAEL, or BMDL for the critical effect in the most appropriate or sensitive mammalian species. This approach is similar to the one used to derive the acute and chronic reference doses (RfD) or Acceptable Daily Intake (ADI) except that it is specific for reproductive and developmental effects and is derived specifically for the exposure duration of concern in the human. The evaluative process uses the UEL both to avoid the connotation that it is the RfD or reference concentration (RfC) value derived by EPA or the ADI derived for food additives by the Food and Drug Administration, both of which consider all types of noncancer toxicity data. Other approaches for more quantitative dose-response evaluations can be used when sufficient data are available. When more extensive data are available (for example, on pharmacokinetics, mechanisms, or biological markers of exposure and effect), one might use more sophisticated quantitative modeling approaches (e.g., a physiologically based pharmacokinetic or pharmacodynamic model) to estimate low levels of risk. Unfortunately, the data sets required for such modeling are rare.

Calculation of the Margin of Exposure (MOE)

The MOE is the ratio of the NOAEL or BMDL to the anticipated human exposure. The higher the ratio, the greater the numerical distance between the human exposure estimate and the dose that is at the lower end of the range of concern from animal studies. The adequacy of the MOE should be evaluated on the same basis as for the UEL. For example, if the NOAEL in a rat developmental toxicology study is 10 mg/kg/d and the human exposure anticipated is 0.1 mg/kg/d, the MOE is 100. The adequacy of the MOE should consider interspecies and intraspecies variability and any other uncertainties accounted for in the UFs applied to derive the UEL. If the MOE is less than the total UF applied in calculating the UEL and is judged to be inadequate, then exposure must be reduced either by applying controls or by removing the exposure.

Because human exposure might differ in different settings, there can be different MOEs for different circumstances. For example, the MOE for an occupational setting might differ from the MOE for environmental exposure. An MOE of 100 is equivalent to exposure at the UEL if the UEL is adjusted by 2 orders of magnitude for uncertainty factors that represent interspecies and intraspecies variability. It should be emphasized, though, that the use of those factors is based on judgment rather than on the default assumption that they are appropriate in all instances. In some cases, a small MOE can be considered protective of health. For example, the lithium assessment conducted using a similar evaluative process (J.A. Moore et al. 1995b) resulted in MOEs of 11-108 for exposures other than the therapeutic use of lithium. Because the size of the total UF applied to the NOAEL was 10, these MOEs were judged to be protective.

Some regulatory agencies use the MOE as an action level. Which MOE is selected is a matter of policy rather than science. For example, California's Proposition 65 relies on an MOE of 1,000 to select exposures to a particular agent for regulatory action. The institution of action based on any given MOE does not mean that toxicity will occur at a lower MOE, only that the chosen MOE is believed to be protective. The subcommittee recommends against using a particular value as an action level because it is arbitrary and does not take into account the complexity and uncertainties inherent in such assessment processes or the variability in differing exposure situations.

As in the case for the choice of UFs and modifying factors in the calculation of an UEL, the choice of an MOE for regulatory action can be based on the level of confidence in the underlying data and on judgment about other factors that might influence risk.

Assessing a Degree of Concern

EPA (1991, 1996a) and the International Programme on Chemical Safety (IPCS 1995) have proposed approaches for characterizing the database concerning potential reproductive and developmental toxicity and provided a basis for how to assess the degree of concern. In its developmental toxicity (EPA 1991) and reproductive toxicity (EPA 1996a) risk-assessment guidelines, EPA uses a weight-of-evidence approach to determine whether a substance poses a risk to humans based on an overall evaluation of reproductive and developmental toxicity and exposure data.

Some aspects of degree of concern currently can be considered in a quantitative evaluation. For example, EPA considers human and animal data in the process of calculating the RfD, and these data are used as the critical effect when they indicate that developmental effects are the most sensitive endpoints. When a complete database is not available, a database UF is recommended to account for inadequate or missing data. The dose-response nature of the data is considered to an extent in the RfD process, especially when the BMD approach is used to model data and to estimate a low level of response; however, there is no approach for including concerns about the slope of the dose-response curve. Because concerns about the slope of the dose-response curve are related to some extent to human exposure estimates, this issue must be considered in risk characterization. (If the MOE is small and the slope of the dose-response curve is very steep, there could be residual uncertainties that must be dealt with to account for the concern that even a small increase in exposure could result in a marked increase in response.) On the other hand, a very shallow slope could be a concern even with a large MOE, because definition of the "true" biological threshold will be more difficult and an additional factor might be needed to ensure that the RfD is below that threshold.

As an example, consider two compounds that are candidates for use in an occupational setting. Both compounds have an MOE of 100, but one is an alkylating agent with a steep dose-response curve. Although the MOEs are the same in this setting, it is reasonable to select the compound that is not an alkylating agent because it is associated with concern at a higher dose. The NOAEL approach does not allow for considerations of variability in the data, but use of the lower confidence limit on dose in the BMD approach does account for variability in the animal or human data on which it is based, even though it does not account for all intraspecies variability.

Both the interspecies and intraspecies UFs include consideration of potential toxicokinetic and toxicodynamic similarities and differences among species and within humans, and those factors can be adjusted when data are available to account more appropriately for similarities and differences among species and within human subpopulations, including different age groups. For example, if it is assumed that a portion of interspecies and intraspecies variability is the result of differences in kinetics, information about the kinetics of a substance in the experimental animal model and in humans might indicate less uncertainty about the extrapolation of effect levels. When there is less uncertainty, lower UFs can be considered. An example of an evaluation in which lower UFs were used is the evaluation of lithium by Moore et al. (1995b). Lower UFs were used because the risk assessment was performed using achieved serum concentrations, rather than administered doses, obviating the influence of absorption and distribution differences among and within species, and because biotransformation of lithium does not occur. Although various means of accounting for degree of concern are described here, there is no formal process for doing so, and this issue should be further considered for the appropriate calculation of the UEL.

Although it is tempting to use default values for UFs or MOEs in the regulation of human exposures, the importance of professional judgment in evaluating the data set for a given substance makes it advisable to avoid inflexible approaches to regulating exposures. Consideration of the reliability of the reproductive and developmental data set and of non-reproductive toxicity data can be an important part of the evaluative process. The inclusion of all available information in the evaluation and the use of scientific judgment are recommended as most likely to lead to the most informed estimate of the risk of anticipated human exposure.

Application of Reproductive and Developmental Toxicity Data to Various Exposure Scenarios

Although the product of the evaluative process is an effect level that can be manipulated in different ways to estimate the risk posed by human exposures, considerable judgment is necessary in the evaluation of the confidence with which the estimation can be performed. There are instances in which the quality of a data set is sufficient to permit the determination of a NOAEL or BMD, but characteristics of the data set detract from the certainty that the NOAEL or BMD will give rise to an appropriate UEL or MOE. A difference between the dosing schedule and route of exposure used in an experimental study and the anticipated human dose and route of exposure are examples of factors that might undermine confidence in the predictive value of the experimental data. Even when the dose and route of exposure are the same as the anticipated human dose and route of exposure, the exposure pattern might be sufficiently different to decrease confidence in extrapolation from the data.

The data available on reproductive and developmental toxicity usually come from studies using repeated dosing regimens that can be characterized as short-term or subchronic exposures. For example, in the prenatal developmental toxicity study, dosing covers a period of development equivalent to the first and part of the second trimester of human gestation. In the developmental neurotoxicity study, the dosing period is both prenatal and early postnatal to cover most of nervous system development. In the two-generation reproduction study, animals are exposed continuously through both generations. The NOAELs and BMDs for all developmental toxicity studies should be compared with all other toxicity data so that, in the cases where the NOAEL or BMD for developmental toxicity is lower than the NOAEL or BMD for chronic toxicity, they can be used as the basis for the UEL and be protective of children's health.

Although there are no developmental studies in which an acute (single) dosing regimen is used to meet regulatory requirements, a central premise in developmental toxicology is that adverse developmental outcomes can result from a single pre- or postnatal exposure. An experimental animal study that uses once-daily gavage might not produce the same exposure profile as human dietary exposure to an agricultural chemical, for example. Although it has been customary to use effect levels from experimental studies without regard to differences in dosing profiles, supplemental information about the activity of a substance might suggest that dosing profile differences should be considered.

If, for example, it is clear that a substance produces toxicity by reaching a peak concentration in the plasma, a single-dose gavage study will be more likely to reach that peak than will administration in the diet. A NOAEL from a single-dose gavage study might be regarded with a different level of confidence under those conditions than would a NOAEL from a dietary study. Data are available to show that most of the types of developmental endpoints from studies used to evaluate pre- and postnatal toxicity (prenatal developmental toxicity, developmental neurotoxicity, two-generation reproduction studies) can result from single exposures. It is recognized that some outcomes might result only from repeated exposure to a given substance and the degree of reversibility of the effect might depend on the duration of exposure. For example, plasma concentrations of agents that induce their own metabolism are lower after a few days of exposure than on the first day of dosing. Thus, a single exposure on a critical day might be of concern. However, some agents might require repeated exposures to reach steady-state plasma concentrations. Therefore, a single-exposure study would underestimate the toxicity of repeated exposures. Determining whether a particular developmental outcome results from a single acute exposure or from repeated exposures requires additional studies that are not often available. Information on toxicokinetics or mechanisms of action might be helpful in interpreting the data but, again, such information is not typically available. As a default, data from all studies that evaluate reproductive and developmental toxicity should be considered in determining UELs for acute, short-term, and longer-term exposure scenarios.

Most developmental toxicity studies (of all types) are conducted using the oral route of exposure. In some cases, dermal exposure is used and, rarely, inhalation exposure. Route-to-route extrapolation is sometimes done to allow consideration of developmental toxicity data. Pharmacokinetic data on different routes of exposure can be extremely useful in the extrapolation of data between routes.

Critical Data Needs

A primary objective of the evaluative process is to use data to formulate and express judgments about reproductive and developmental risk potential for humans. Flawed or nonexistent data compromise the certainty of scientific judgment. Although guidance or regulations promulgated by government agencies serve definite needs, they are somewhat rigid. For the evaluative process that is proposed here, it seems best to determine the adequacy of the database in a case-specific manner.

It is better to ascertain toxicity and estimate dosimetry using a species in which the metabolism pathway for that agent parallels that of humans than it is to try to assess toxicity and dosimetry in two randomly selected species in which metabolism of the substance is either unknown or is dissimilar to that of humans. For example, methanol, which is acutely toxic to humans and nonhuman primates, is metabolized via a folate-dependent pathway. However, rodents use a different folate-dependent pathway, and the rate at which rodents detoxify formate (a metabolite of methanol and the agent that causes toxicity to humans exposed to high doses) is more rapid than that in primates. That increased rate can be attributed to higher levels of hepatic tetrahydrofolate, the enzyme responsible for the oxidation of formate (Tephly and McMartin 1984; Johlin et al. 1987; Medinsky and Dorman 1995). Therefore, rodent species would not be the best to use for extrapolating acute, high-dose methanol toxicity in humans. On the other hand, for human exposures to methanol at concentrations below the threshold for formate accumulation, rodent models can be useful because they provide the advantage of allowing dose-response studies in which the animals do not experience formate build-up (Rogers et al. 1993).

During the review of existing information, evaluators might identify some data as insufficient for judging human risks, either because the data do not exist or because they are compromised in some way for risk assessment. In another chemical evaluation, data might be judged sufficient to determine human risk potential, but in the judgment of the evaluators there might be large degrees of uncertainty because of reliance on default assumptions or because of the inherent uncertainty in some of the data that are central to the evaluation. In each instance, evaluators will cite specific data needs if they determine that the data will materially improve the certainty of an existing judgment about human risk.

SUMMARY

The selection of the term "summary," instead of "risk characterization," to describe the concluding step in the evaluative process is deliberate. The evaluative process described here focuses on reproductive and developmental effects; it does not account for all the effects that should be included in a risk characterization. Moreover, a lack of detailed exposure information will be common in this type of evaluative process but is necessary for a risk characterization.

In this case, the summary communicates to Navy environmental health practitioners scientific judgment on chemical risk for reproductive and developmental toxicity. The degree of certainty of the judgment must be expressed in terms that are meaningful to those with a general science background in toxicology and risk assessment. The key to achieving that goal is candor in explaining the basis of the judgment, its breadth of support, and, especially, the degree to which the judgment reflects actual information, confident extensions from closely related data, or the invoking of assumptions when no information is available.

The summary is written from statements developed in the integrated evaluation and quantitative assessment steps of the evaluative process. The summary will review the following elements:

- *Background.* This section provides a brief, readable review of the general chemical, toxicological, and biological characteristics of the substance.
- *Human exposure.* This section gives a clear statement of the conditions of use or ambient concentrations that might produce different doses, routes, or frequencies of human exposure. It describes how different patterns of use produce differences in the magnitude of exposure.
- *Toxicology.* Summaries of developmental toxicity, and of male and female reproductive toxicity, appear in this section. The discussion also contains statements about the sufficiency and relevance of the data.
- *Quantitative evaluation.* This section lists the quantitative values derived in the evaluative process and states the degree to

which the values are derived from actual data or reflect the use of default assumptions.

- *Certainty of judgment and data needs.* The use of default assumptions, while often necessary, represents a tangible expression of uncertainty. To clarify that point, this section discusses the magnitude of an assumption's influence on the judgments made in the evaluation. Where the effect is large and the uncertainty great, the evaluators might sometimes defer a judgment. Where a default assumption has a major effect on the evaluative judgment, the evaluative summary clearly defines the kind of data needed to supplant the default and identities that as a critical data need.

  Only some aspects of the assessment might involve uncertainty of judgment. For example, although there might be great certainty that the data qualitatively predict human health risk potential, the nature and degree of exposure might be poorly understood. In that case, the evaluative summary will clearly state that there is reasonable certainty of human risk potential and explain why the quantitative uncertainty (missing, inadequate exposure data) leads to the use of a conservative default assumption that is likely to overestimate the degree of exposure and risk.

- *References.* In any evaluation of this nature, a bibliography is imperative. All literature reviewed should appear in a reference list. A separate listing of references reviewed but not used in the evaluation also should appear in the document.

4

Incomplete or Insufficient Data Sets

The evaluative process described in Chapters 2 and 3 and illustrated in Appendix A for 1,1,1,2-tetrafluoroethane (HFC-134a) and jet fuel JP-8 permits classification of exposure to a substance with respect to its reproductive and developmental toxicity *if* sufficient scientific data are available. The database required in each instance to permit confident classification must adequately characterize the full range of reproductive and developmental toxicity in humans and evaluate the associated hazards. In addition, the actual range of conditions of exposure must be known in sufficient detail to determine whether the dose, duration, timing, route, and other characteristics of exposure pose a substantial reproductive risk. In practice, such a complete database is rarely available. As will be apparent in Appendix A, the databases for JP-8 and HFC-134a are sufficient only in some aspects. This chapter discusses a common-sense approach to minimizing the risk when there are insufficient data available regarding a particular exposure to permit confident determination of the associated reproductive and developmental toxicity. This is a further application of the approach explained in Chapters 2 and 3 for interpreting toxicity data. The rationale is to avoid circumstances that create a high degree of concern.

PRINCIPLES TO MINIMIZE RISK

The principles used to minimize risk are the same whether or not the risk can be fully characterized. The principles also are the same whether the risk under consideration is one related to developmental toxicity, reproductive toxicity, or another type of toxicity (e.g., pulmonary, neurological, renal).

The toxicity produced by a particular exposure depends on a variety of factors, including the following:

- The chemical and toxicological nature of the agent itself.
- The agent's physical properties (e.g., solubility, volatility).
- The conditions of exposure (e.g., dose, duration, frequency, timing, route).
- The use of safety measures that reduce actual exposure (e.g., gloves, masks, and ventilation).
- The agent's pharmacokinetics of absorption, metabolism, distribution, and excretion, all of which are subject to individual variability.
- The pharmacodynamics (target organ, site of action, receptor interactions) that determine the agent's mechanism of action.
- Concomitant exposures to other chemical or physical agents that affect the factors listed above or the agent's toxic activity.
- Biological characteristics of the exposed individual (e.g., pregnancy, age, nutritional status, genetic susceptibility).

The only way to eliminate completely the possibility of toxic effects associated with exposure to an agent is to eliminate all exposures. When use of the agent is necessary, minimizing the exposure will minimize risk.

The guidelines below are for exposures that have not been adequately characterized with respect to reproductive and developmental toxicity, but for which there may be other data on toxicity, as listed above.

- Use expert judgment to evaluate the available toxicity data.
- Assume that susceptibility to reproductive or developmental toxicity may be greater than susceptibility to any known toxicity of the agent, and apply additional uncertainty factors to reflect the lack of data.
- Substitute exposure to an agent that is known not to be associated with substantial risk for causing reproductive and developmental toxicity for an agent associated with unknown risk.

- If use of a given agent is unavoidable, the risk should be minimized by limiting the potentially absorbed dose.

PRACTICAL APPLICATION

When the reproductive and developmental risks of a particular exposure cannot be fully characterized, a conservative approach is to assume that such risks exist with exposure conditions below those that produce toxicity for the most sensitive system known for the agent. Toxicity to the "most sensitive system" is that produced by exposure to the agent at the lowest effective dose in any relevant species studied.

The confidence with which one can apply this assumption to a particular exposure that has not been adequately studied depends on the amount of relevant information about the exposure in general. The less that is known, the greater the uncertainty, and the greater the degree of concern. This means there must be an inverse relationship between the quality of the information regarding reproductive and developmental toxicity and the size of the uncertainty factor required beyond the exposure limit, based on other kinds of toxicity. In general, however, the uncertainty factor should produce an exposure limit that is lower than that based on known toxicity of other kinds when the database for reproductive and developmental toxicity is inadequate.

No chemical substance should be misused or abused. Whenever possible, an exposure that is known not to be associated with substantial reproductive or developmental toxicity should be substituted for an exposure associated with unknown risks. The decision to substitute one exposure for another or to eliminate a particular exposure altogether is always complicated. The factors to consider include physical and chemical characteristics such as molecular weight, volatility and vapor pressure, and the octanol-water partition coefficient (Kow). For example, if a chemical is a polymer with a molecular weight high enough that it is unlikely to be absorbed or distributed, it may be excluded from concern. Additional factors include other kinds of toxicity, ease and safety of storage, availability, and cost. Although reproductive and developmental toxicity can never be the only issues considered in such decisions, they must never be ignored.

Reproductive and developmental toxicity are less likely to be associated with exposures that minimize absorption of the agent than with greater exposures. Exposures associated with an unknown risk of reproductive and developmental toxicity should be as brief and as infrequent as possible. The amount of the agent used should be restricted and the method of use should be designed to reduce possible absorption. In addition, reasonable safety procedures and engineering safeguards should limit the exposure. Volatile materials should be used with adequate ventilation, if possible, and with appropriate masks or respirators when adequate ventilation is not possible. Proper gloves and protective garments should be used when handling materials that might be absorbed through the skin. Appropriate washing procedures should be used, and eating, drinking, and smoking should be prohibited in circumstances that might permit inadvertent ingestion of the agent.

REDUCING UNCERTAINTY

The approach described in this chapter results in limitations that would be unnecessary if the exposure under consideration were known to be safe. The most effective way to reduce uncertainty is to develop data to characterize the toxicity of the agents the Navy must use. Applying additional uncertainty factors to exposure limits can lead to unnecessarily conservative limits, which can lead the Navy to curtail the use of agents that may have been acceptable if adequate data were available. Such circumstances may increase costs. Obtaining a sufficient data set on the safety of a particular exposure might, therefore, provide substantial savings without increasing the risk of reproductive or developmental toxicity.

Application of this approach will highlight those substances for which obtaining a more complete data set is a high priority. The Subcommittee on Reproductive and Developmental Toxicology recommends that the Navy commission or undertake the studies necessary to obtain additional information on agents in use or intended for use for which there is little or no information as an important means of reducing both uncertainty and expense.

The subcommittee also recommends developing improved exposure assessments unique to the Navy environment and pertinent to exposures that are particularly important to assessing reproductive and developmental toxicity. Those include exposures to males and females alike in various occupational situations and accounting for such factors as body size, the differences in the nature of the work and the workplace, and other factors that affect exposure, as well as the potential for contamination of the home environment secondary to workplace exposures.

5

Recommendations

Chapters 2, 3, and 4 of this report describe the process recommended by the Subcommittee on Reproductive and Developmental Toxicology for evaluating exposures to agents for reproductive and developmental toxicity. Appendix A illustrates how that process can be used. This chapter contains a discussion of several general recommendations and areas of research that the subcommittee believes would improve the Navy's ability to evaluate exposures to agents.

GENERAL RECOMMENDATIONS

- *Agents should not be classified simply as toxic or nontoxic to reproduction and development; rather, potential risks should always be considered in the context of exposure.*

The risk of adverse reproductive or developmental effects from exposure to a given substance should be considered only in the context of the exposure situation. In this way, both the agent itself and the conditions of exposure, including the dose, route, timing, and duration of exposure are considered, rather than "labeling" an agent either as "toxic" or as "safe."

The subcommittee acknowledges that the Navy might need to use a screening process in which decisions are made in a dichotomous manner (to use or not to use a particular agent). Such decisions can be made by considering the exposure scenario that is anticipated in the workplace. The evaluative process describes an approach by which an exposure level that is unlikely to be associated with reproductive and developmental toxicity (the unlikely effect level; UEL) can be estimated. If the workplace scenario is anticipated to result in human exposures sufficiently lower than that estimate, then for policy decisions, the exposure can be regarded as acceptable. If the anticipated human exposure is higher than the estimated UEL, the use of the agent in question can be regarded as unacceptable, and alternative agents can be evaluated or exposure control measures can be put into place.

- *The evaluative process should be implemented by a team of scientists with training and experience in assessing reproductive and developmental toxicity.*

The process described by the subcommittee requires expertise in the intricacies and relationships of several integrated processes involved in reproduction and development and the exercise of considerable judgment based on the body of scientific knowledge in these areas. That judgment is brought to bear in interpreting data and making decisions concerning the adequacy of available data sets for estimating the potential reproductive and developmental toxicity of agents under specific conditions of exposure. In addition, once there has been a determination of the exposure at which adverse effects are unlikely, judgment is required in the evaluation of other characteristics of the agent or exposure conditions that might make it advisable to alter the estimate for a given workplace scenario.

- *In cases in which the data set is incomplete or insufficient, evaluators should assume that susceptibility to reproductive or developmental toxicity may be greater than susceptibility to any known toxicity of the agent, and apply additional uncertainty factors to reflect the degree of uncertainty attributable to missing data.*

When the reproductive and developmental risks of a particular exposure cannot be fully characterized, a conservative approach is to assume that such risks exist with exposure conditions below those that produce toxicity for the most sensitive system known for the agent.

The most sensitive system is the kind of toxicity produced by exposure to the agent at the lowest effective dose in any relevant species studied.

The confidence with which one can apply this assumption to a particular exposure that has not been adequately studied depends on the amount of relevant information available about the exposure in general. The less that is known, the greater the uncertainty, and the greater the degree of concern. When the database for reproductive and developmental toxicity is inadequate, the total uncertainty factor applied should result in an exposure limit that is lower than that based on known toxicity of other kinds.

- *An exposure to an agent that is known not to be associated with substantial reproductive or developmental toxicity should be substituted for an exposure to an agent associated with unknown risks.*

The decision to substitute one agent for another or to eliminate a particular agent altogether always involves many factors, including other kinds of toxicity, ease and safety of storage, availability, and cost. Although reproductive and developmental toxicity can never be the only issue considered in such decisions, it must never be ignored.

- *When use of an agent with demonstrated toxicity is necessary, minimize the potential risk by limiting the potentially absorbed dose.*

Reproductive and developmental toxicity is less likely to be associated with exposures that minimize absorption of the agent. Exposures associated with an unknown risk of reproductive and developmental toxicity should be as brief and as infrequent as possible. The amount of an agent used should be minimized, and reasonable safety procedures and engineering safeguards should be used to limit the exposure.

- *The Navy should consider using the National Library of Medicine's Developmental and Reproductive Toxicology (DART) database as the primary source of bibliographic information in this area.*

DART is a bibliographic database that covers the literature on teratology and other aspects of reproductive and developmental toxicology. DART is an essential resource to the Navy for gathering information on the potential reproductive and developmental effects of agents because it greatly simplifies the process for searching for literature in this area.

In addition to DART, there are a number of additional sources of information that the Navy should consider using to evaluate the reproductive and developmental toxicity potential of agents. These sources are described in Appendix B.

RESEARCH RECOMMENDATIONS

- *The Navy should conduct or commission studies that are necessary to obtain sufficient data sets for the agents it is considering for use.*

As noted in Chapter 1, for a data set to be considered sufficient, it should include consideration of potential adverse reproductive and developmental effects of male and female exposure. The absence or inadequacy of data on one or more of the components of reproductive toxicity (male reproductive effects, female reproductive effects, developmental effects) does not equate with lack of effect.

Data on the reproductive and developmental toxicity of exposures to agents are often sparse and, when data are available, there can be variability in the quality of the studies from which they are obtained. To account for such incomplete or inadequate data sets, an uncertainty factor is applied to the no-observed-adverse-effect level, lowest-observed-adverse-effect level, or benchmark dose. It is possible that such uncertainty could lead to the calculation of an exposure limit that is more conservative than necessary and, based on those limits, the Navy could decide to curtail use of specific substances or institute costly exposure control measures. Uncertainty could be reduced by filling in data gaps and improving exposure estimates so that the potential toxicity of agents that the Navy is considering for use is better understood. Developing a sufficient data set on a particular exposure to an agent could provide a cost savings to the Navy and reduce the risk of reproductive and developmental toxicity.

Because there are exposure scenarios that are unique to the Navy's work environment, the subcommittee also recommends that the Navy consider developing a research program to meet needs that are not being met by civilian research (e.g., the study of reproductive and developmental toxicity in the context of naval operations). Such a program would allow the Navy to anticipate and rank the agents it would like to use and to study the reproductive and developmental toxicity of those agents before its personnel are exposed.

- *The Navy should monitor Navy personnel for reproductive and developmental outcomes.*

The Navy is well equipped to design, implement, and conduct epidemiological studies that focus on various reproductive and developmental outcomes. Such studies should include male and female military and civilian personnel as well as other populations at risk (e.g., partners of naval personnel and residents of communities affected by naval operations). Naval ships provide a unique opportunity to study a well-defined population, and one in which many confounders that affect community or occupational studies (e.g., lifestyle factors thought to affect reproductive health such as alcohol consumption and cigarette smoking) can be documented. As such, the Navy is well suited to conduct surveillance, record linkage, and etiological studies.

Specifically, the following activities could be conducted:

- A complete and up-to-date reproductive history should be obtained and available for all naval personnel (men and women; active duty and reservists). Such a history should be updated annually or after a reproductive outcome. This would provide important baseline information and permit study of maternally and paternally mediated effects. The reproductive history should address sexual activity and inactivity, sexual libido, sexual dysfunction, semen analysis, menstruation history, pregnancy intentions, time-to-pregnancy (conception delays, fecundability, infertility), and pregnancy outcomes (e.g., ectopic pregnancy, spontaneous loss, fetal demise, birth size, secondary sex ratios, birth defects, mental retardation, developmental disabilities). Recording this information is in keeping with the definition for reproductive health and the need to address all health aspects of individuals.
- Surveillance of reproductive health could be obtained via record linkages with live birth or fetal death registeries or via the establishment of outcome-specific registries (such as a congenital malformation registry for the navy). Baseline prevalence figures for various outcomes are urgently needed for military populations; the U.S. general population might not be an appropriate reference group.
- Hypothesis-driven etiological studies can be designed on an ad hoc basis in response to concerns or based on associations

observed in linkage or surveillance studies. The studies should be designed to address questions of utmost concern, and they should be grounded within the epidemiological method or framework for study.

- *The Navy should conduct toxicity studies on chemical mixtures.*

In many cases, exposure is not to an individual chemical, but to mixtures of chemicals. Risks to human health from multichemical exposures often are not well understood. Because multichemical exposures are found in the Navy workplace, the Navy should conduct research on the toxicity of these mixtures. Research also should be conducted to illuminate exposure scenarios associated with chemical mixtures.

References

ACGIH (American Conference of Governmental Industrial Hygienists). 2000. 2000 TLVs and BEIs. Publication 0100. American Conference of Governmental Industrial Hygienists, Cincinnati, OH.

Agnish, N.D., and K.A. Keller. 1997. The rationale for culling of rodent litters. Fundam. Appl. Toxicol. 38(1):2-6.

AIHA (American Industrial Hygiene Association). 1991. Workplace Environmental Exposure Level Guide. 1,1,1,2-Tetrafluoroethane. American Industrial Hygiene Association, Akron, OH.

Allen, B.C., R.J. Kavlock, C.A. Kimmel, and E.M. Faustman. 1994a. Dose-response assessment for developmental toxicity: II. Comparison of generic benchmark dose estimates with no observed adverse effect levels. Fundam. Appl. Toxicol. 23(4):487-495.

Allen, B.C., R.J. Kavlock, C.A. Kimmel, and E.M. Faustman. 1994b. Dose-response assessment for developmental toxicity: III. Statistical models. Fundam. Appl. Toxicol. 23(4):496-509.

Alexander, D.J., and S.E. Libretto. 1995. An overview of the toxicology of HFA-134a (1,1,1,2-tetrafluoroethane). Hum. Exp. Toxicol. 14(9):715-720.

Alexander, D.J., S.E. Libretto, M.J. Adams, E.W. Hughes, and M. Bannerman. 1996. HFA-134a (1,1,1,2-tetrafluoroethane): Effects of inhalation exposure upon reproductive performance, development and maturation of rats. Hum. Exp. Toxicol. 15(6):508-517.

Andersen, M.E., M.E. Meek, G.A. Boorman, D.J. Brusick, S.M. Cohen, Y.P. Dragan, C.B. Frederick, J.I. Goodman, G.C. Hard, E.J. O'Flaherty, and D.E.

Robinson. 2000. Lessons learned in applying the U.S. EPA proposed cancer guidelines to specific compounds. Toxicol. Sci. 53(2):159-172.

Anderson, D. and C.R. Richardson. 1979. Arcton-134a: A Cytogenetic Study in the Rat. ICI Rep. NO. CTL/P/444. Imperial Chemical Industries, Alderley Park, Macclesfield, Cheshire, U.K.

Armstrong, B.K., E. White, and R. Saracci. 1995. Principles of Exposure Measurement in Epidemiology. Monographs in Epidemiology and Biostatistics, Vol 21. New York: Oxford University Press.

ATSDR (Agency for Toxic Substances and Disease Registry). 1998. Toxicological Profile for JP-5 and JP-8. Prepared by Research Triangle Institute, Contract 205-93-0606, for the U.S. Department of Health and Human Services, Agency for Toxic Substances and Disease Registry, Atlanta, GA.

Azar, A., H.J. Trochimowicz, J.B. Terrill, and L.S. Mullin. 1973. Blood levels of fluorocarbon related to cardiac sensitization. Am. Ind. Hyg. Assoc. J. 34(3):102-109.

Bagur, A.C., and C.A. Mautalen. 1992. Risk for developing osteoporosis in untreated premature menopause. Calcif. Tissue Int. 51(1):4-7.

Barlow, S.M., and F.M. Sullivan. 1982. Reproductive Hazards of Industrial Chemicals: An Evaluation of Animal and Human Data. New York: Academic Press.

Barnes, D.G., G.P. Daston, J.S. Evans, A.M. Jarabek, R.J. Kavlock, C.A. Kimmel, C. Park, and H.L. Spitzer. 1995. Benchmark dose workshop: Criteria for use of a benchmark dose to estimate a reference dose. Regul. Toxicol. Pharmacol. 21 (2):296-306.

Barone, S. Jr., M.E. Stanton, and W.R. Mundy. 1995. Neurotoxic effects of neonatal triethyltin (TET) exposure are exacerbated with aging. Neurobiol. Aging. 16(5):723-735.

Barr, M. 1997. Lessons from human teratogens: ACE inhibitors. Teratology 56(6): 373.

Barton, S.J., P. McDonald, and J. Sandow. 1994. HFA-134a: Study of the Effects on Testicular Endocrine Function After Inhalation Exposure (6 h per day). IRI Rep. No. 7955. Inveresk Research International, Tranent, Scotland.

Bogdanffy, M.S., G. Daston, E. M. Faustman, C. A. Kimmel, G. L. Kimmel, J. Seed, and V. Vu. In press. Harmonization of cancer and non-cancer risk assessment: Proceedings of a consensus-building workshop. Toxicol. Sci.

Bogo, V., R.W. Young, T.A. Hill, R.M. Cartledge, J. Nold, and G.A. Parker. 1983. The Toxicity of Petroleum and Shale JP-5. Proceedings of the First Toxicology of Petroleum Hydrocarbons Symposium, Armed Forces Radiobiology Institute, Bethesda, MD.

Boyle, C.A., P. Decoufle, and M. Yeargin-Allsopp. 1994. Prevalence and health impact of developmental disabilities in US children. Pediatrics 93(3):399-403.

Brown, J., III, B. Burke, and A.S. Dajani. 1974. Experimental kerosene pneumonia: Evaluation of some therapeutic regimens. J. Pediatr. 84(3):396-401.

Bruner, R.H. 1984. Pathologic findings in laboratory animals exposed to hydrocarbon fuels of military interest. Pp. 133-140 in Advances in Modern Environmental Toxicology, Vol VII: Renal Effects of Petroleum Hydrocarbons, M.A. Mehlman, G.P Hemstreet III, J.J. Thorpe, and N.K. Weaver, eds. Princeton, NJ: Princeton Scientific Publishers.

Bruner, R.H., E.R. Kinkead, T.P. O'Neill, C.D. Flemming, D.R. Mattie, C.A. Russell, and H.G. Wall. 1993. The toxicologic and oncogenic potential of JP-4 jet fuel vapors in rats and mice: 12-Month intermittent inhalation exposures. Fundam. Appl. Toxicol. 20:97-110.

California Environmental Protection Agency. 1991. Draft Guidelines for Hazard Identification and Dose-Response Assessment of Agents Causing Developmental and/or Reproductive Toxicity. California Department of Health Services, Health Hazard Assessment Division, Reproductive and Cancer Hazard Assessment Section. April 3.

Callander, R.D., and K.P. Priestley. 1990. HFC-134a: An Evaluation Using the Salmonella Mutagenicity Assay. ICI Rep. No. CTL/P/2422. Central Toxicology Laboratory. Imperial Chemical Industries, Alderley Park, Macclesfield, Cheshire, U.K.

Carlton, G.N., and L.B. Smith. 2000. Exposures to jet fuel and benzene during aircraft fuel tank repair in the U.S. Air Force. Appl. Occup. Environ. Hyg. 15(6):485-491.

Carpenter, C.P., D.L. Geary, Jr., R.C. Myers, D.J. Nachreiner, L.J. Sullivan, and J.M. King. 1976. Petroleum hydrocarbon toxicity studies: XI. Animal and human response to vapors of deodorized kerosene. Toxicol Appl. Pharmacol. 36(3): 443-456.

Casaco, A., R. Gonzalez, L. Arruzazabala, M. Garcia, and A.R. de la Vega. 1982. Studies on the effects of kerosine aerosol on airways of rabbits. Allergol. Immunopathol. 10(5):361-366.

CDC (Centers for Disease Control and Prevention). 1995. Economic costs of birth defects and cerebral palsy — United States, 1992. MMWR 44:694-699.

Chapin, R.E., S.L. Dutton, M.D. Ross, and J. Lamb, IV. 1985. Effects of ethylene glycol monomethyl ether (EGME) on mating performance and epididymal sperm parameters in F344 rats. Fundam. Appl. Toxicol. 5:182-189.

Chen, H., M.L. Witten, J.K. Pfaff, R.C. Lantz, and D. Carter. 1992. JP-8 jet fuel exposure increases alveolar epithelial permeability in rats [abstract]. FASEB J. 6(4):A1064.

Christian, M.S. 1986. A critical review of multigeneration studies. J. Am. Coll. Toxicol. 5(2):161-180.

Clarke, D.O., J.M. Duignan, and F. Welsch. 1992. 2-Methoxyacetic acid dosimetry- teratogenicity relationships in CD-1 mice exposed to 2-methoxyethanol. Toxicol. Appl. Pharmacol. 114(1):77-87.

Clegg, E.D., J.C. Cook, R.E. Chapin, P.M. Foster, and G.P. Daston. 1997. Leydig cell hyperplasia and adenoma formation: Mechanisms and relevance to humans. Reprod. Toxicol. 11(1):107-121.

Collins, M.A. 1984. HFC134a: Acute toxicity in rats to tetrafluroethane. Central Toxicology Laboratory, Imperial Chemical Industries, Alderley Park, Macclesfield, Cheshire, U.K.

Collins, M.A, G.M. Rusch, F. Sato, P.M. Hext, and R.-J. Millischer. 1995. 1,1,1,2-Tetrafluoroethane: Repeat exposure inhalation toxicity in the rat, developmental toxicity in the rabbit, and genotoxicity in vitro and in vivo. Fundam. Appl. Toxicol. 25:271-280.

Cooper, J.R., and D.R. Mattie. 1996. Developmental toxicity of JP-8 jet fuel in the rat. J. Appl. Toxicol. 16(3):197-200.

Cowan, M.J., Jr. and L.J. Jenkins, Jr. 1981a. U.S. Navy toxicity study of shale and petroleum JP-5 aviation fuel and marine diesel fuel. Pp. 129-140 in Health Effects Investigation of Oil Shale Development, W.H. Griest, M.R. Guerin, and D.L. Coffin, eds. Ann Arbor, MI: Ann Arbor Science Publishers.

Cowan, M.J., and L.J. Jenkins. 1981b. The Toxicity of Grade JP-5 Aviation Turbine Fuel, A Comparison Between Petroleum and Shale-Derived Fuels. Pp. B2/1-B2/7 in Toxic Hazards in Aviation. Papers presented at the Aerospace Medical Panel Specialists' Meeting, Toronto, Canada, Sept. 15-19 1980. AGARD-CP-309, France: Advisory Group for Aerospace Research & Development.

Crump, K.S. 1984. A new method for determining allowable daily intakes. Fundam. Appl. Toxicol. 4(5):854-871.

Crump, K.S. 1995. Calculation of benchmark doses from continuous data. Risk Anal. 15(1):79-89.

Daston, G.P., G.J. Overmann, D. Baines, M.W. Taubeneck, L.D. Lehman-McKeeman, J.M. Rogers, and C.L. Keen. 1994. Altered Zn status by alpha-hederin in the pregnant rat and its relationship to adverse developmental outcome. Reprod. Toxicol. 8(1):15-24.

Deichmann, W.B., K.V. Kitzmiller, S. Witherup, and R. Johansmann. 1944. Kerosene intoxication. Ann. Int. Med. 21(Nov.):803-823.

Dourson, M.L., S.P. Felter, and D. Robinson. 1996. Evolution of science-based uncertainty factors in noncancer risk assessment. Regul. Toxicol. Pharmacol. 24(2):108-120.

Dunnett, C.W. 1955. A multiple comparison procedure for comparing several treatments with a control. J. Am. Stat. Assoc. 50:1096-1121.

Dunnett, C.W. 1964. New tables for multiple comparisons with a control. Biometrics 10(Sept.):482-491.

Easley, J.R., J.M. Holland, L.C. Gipson, and M.J. Whitaker. 1982. Renal toxicity of middle distillates of shale oil and petroleum in mice. Toxicol. Appl. Pharmacol. 65(1):84-91.

EC (European Commission). 1992. 7th Amendment: Toxic to Reproduction, Guidance to Classification. Tech. Rep. 47. ISSN-0773-8072-47. European Commission, Brussels, Belgium. August.

ECETOC (European Centre for Ecotoxicology and Toxicology of Chemicals). 1995. Joint Assessment of Commodity Chemicals No. 31. 1,1,1,2-Tetrafluoroethane. (HFC-134a) CAS No. 811-97-2. ISSN-0773-6339-31. European Centre for Ecotoxicology and Toxicology of Chemicals, Brussels, Belgium. February.

Ellis, M.K., L.A. Gowans, T.Green, and R.J.N. Tanner. 1993. Metabolic fate and disposition of 1,1,1,2-tetrafluoroethane(HCFC 134a) in the rat following a single exposure by inhalation. Xenobiotica 23(7):719-729.

Emmen, H.H., and E.M.G. Hoogendijk. 1999. Report on an Ascending Dose Safety Study Comparing HFA-134a with CFC-12 and Air, Administered by Whole-Body Exposure to Healthy Volunteers. TNO Report V98.754-Vol.1-2. Zeist, The Netherlands: TNO Nutrition and Food Research Institute.

EPA (U.S. Environmental Protection Agency). 1987. Pesticide Assessment Guidelines. Subdivision U. Applicator Exposure Monitoring. EPA 540/9-87-127. Office of Pesticide Programs, U.S. Environmental Protection Agency, Washington, DC.

EPA (U.S. Environmental Protection Agency). 1990. Federal Insecticide, Fungicide and Rodenticide Act. (FIFRA): Good Laboratory Practice Standards. U.S. Environmental Protection Agency, Washington, DC.

EPA (U.S. Environmental Protection Agency). 1991. Guidelines for Developmental Toxicity Risk Assessment. Fed. Regist. 56(234):63797-63826.

EPA (U.S. Environmental Protection Agency). 1992. Guidelines for Exposure Assessment. Fed. Regist. 57(104):22888-22938.

EPA (U.S. Environmental Protection Agency). 1994. Methods for Derivation of Inhalation Reference Concentrations and Application of Inhalation Dosimetry. EPA/600/8-90/066F. Office of Research and Development, U.S. Environmental Protection Agency, Washington, DC.

EPA (U.S. Environmental Protection Agency). 1995. The Use of the Benchmark Dose Approach in Health Risk Assessment. EPA/630/R-94/007. Office of Research and Development, U.S. Environmental Protection Agency, Washington DC.

EPA (U.S. Environmental Protection Agency). 1996a. Reproductive Toxicity

Risk Assessment Guidelines. Fed. Regist. 61(212):56273-56322.

EPA (U.S. Environmental Protection Agency). 1996b. Proposed Guidelines for Carcinogen Risk Assessment. EPA/600/P-92/003C. Office of Research and Development, U.S. Environmental Protection Agency, Washington, DC.

EPA (U.S. Environmental Protection Agency). 1997a. Exposure Factors Handbook. Vol. 1. General Factors. EPA/600/P-95/002Fa. Vol.2. Food Integration Factors. EPA/600/P-95/002Fb. Vol.3. Activity Factors. EPA/600/P-95/002Fc. Office of Research and Development, U.S. Environmental Protection Agency, Washington, DC. [Online]. Available: http://www.epa.gov/nceawww1/exposfac.htm. NTIS Publ. Nos. PB98-124225 (Vol. 1); PB98-124233 (Vol. 2); PB98-124241 (Vol. 3); PB98-124217 (Vols. 1-3).

EPA (U.S. Environmental Protection Agency). 1997b. Standard Operating Procedures (SOPs) for Residential Exposure Assessments. Office of Pesticide Programs. U.S. Environmental Protection Agency, Washington, DC. [Online]. Available: http://www.epa.gov/pesticides/SAP/1997/september/1sess3.htm.

EPA (U.S. Environmental Protection Agency). 1998a. Health Effects Test Guidelines OPPTS 870.3700 Prenatal Developmental Toxicity Study. EPA 712-C-98-207. Office of Prevention, Pesticides, and Toxic Substances, U.S. Environmental Protection Agency, Washington, DC. [Online]. Available: http://www.epa.gov/opptsfrs/OPPTS_Harmonized/870_Health_Effe cts_Test_Guidelines/Series/870-3700.pdf

EPA (U.S. Environmental Protection Agency). 1998b. Health Effects Test Guidelines OPPTS 870.3800 Reproduction and Fertility Effects. EPA 712-C-98-208. Office of Prevention, Pesticides and Toxic Substances, U.S. Environmental Protection Agency. Washington, DC [Online]. Available: http://www.epa.gov/opptsfrs/OPPTS_Harmonized/870_Health_Effe cts_Test_G uidelines/Series/870-3800.pdf

EPA (U.S. Environmental Protection Agency). 1998c. Health Effects Test Guidelines OPPTS 870.6300 Developmental Neurotoxicity Study. EPA 712-C-98-239. Office of Prevention, Pesticides and Toxic Substances, U.S. Environmental Protection Agency. Washington, DC [Online]. Available: http://www.epa.gov/opptsfrs/OPPTS_Harmonized/870_Health_Effe cts_Test_G uidelines/Series/870-6300.pdf

EPA (U.S. Environmental Protection Agency). 1998d. Health Effects Test Guidelines OPPTS 870.5450 Rodent Dominant Lethal Assay. EPA 712-C-98-227. Office of Prevention, Pesticides and Toxic Substances, U.S. Environmental Protection Agency. [Online]. Available: http://www.epa.gov/opptsfrs/OPPTS_Harmonized/870_Health_Effects_Test_Guidelines/Series/870-5450.pdf

EPA (U.S. Environmental Protection Agency). 1999. Draft. Toxicology Data Requirements for Assessing Risks of Pesticide Exposure to Children's Health. Report of the Toxicology Working Group of the 10X Task Force, U.S. Environmental Protection Agency, Washington, DC. April 28. [Online]. Available: http://www.epa.gov/scipoly/sap/1999/index.htm#may.

Faustman, E.M., B.C. Allen, R.J. Kavlock, and C.A. Kimmel. 1994. Dose-response assessment for developmental toxicity. I. Characterization of database and determination of no observed adverse effect levels. Fundam. Appl. Toxicol. 23(4):478-486.

FDA (U.S. Food and Drug Administration). 1987. Good Laboratory Practice Regulations for Nonclinical Laboratory Studies, Regulatory Program 21CFR 58. U.S. Food and Drug Administration, Washington, DC.

FDA (U.S. Food and Drug Administration). 1994. International Conference on Harmonisation: Guideline on Detection of Toxicity to Reproduction for Medicinal Products; Availability. Fed. Regist. 59(183):48746-48752.

FDA (U.S. Food and Drug Administration). 2000. Toxicological Principles for the Safety of Food Ingredients. Redbook 2000. Center for Food Safety and Applied Nutrition, U.S. Food and Drug Administration, Washington, DC. [Online]. Available: http://vm.cfsan.fda.gov/~redbook/red-toct.html

Fort, D.J., E.L. Stover, J.A. Bantle, J.R. Rayburn, M.A. Hull, R.A. Finch, D.T. Burton, S.D. Turley, D.A. Dawson, G. Linder, D. Buchwalter, M. Kumsher-King, and A.M. Gaudet-Hull. 1998. Phase III interlaboratory study of FETAX, Part 2: Interlaboratory validation of an exogenous metabolic activation system for frog embryo teratogenesis assay — Xenopus (FETAX). Drug Chem. Toxicol. 21(1):1-14.

Foster, P.M.D. 1989. m-Dinitrobenzene: Studies on its toxicity to the testicular sertoli cell. Arch. Toxicol. (Suppl. 13):3-17.

Francis, E.Z., and G.L. Kimmel. 1988. Proceedings of the Workshop on One-versus Two-Generation Reproductive Effects Studies. J. Am. Coll. Toxicol. 7(7):911-927.

Francis, E.Z., C.A. Kimmel, and D.C. Rees. 1990. Workshop on the qualitative and quantitative comparability of human and animal developmental neurotoxicity: Summary and implications. Neurotoxicol. Teratol. 12(3):285-92.

Friedman, J.M., and J.E. Polifka. 2000. Teratogenic Effects of Drugs: A Resource for Clinicians (TERIS). Baltimore, MD: Johns Hopkins University Press.

Gad, S.C. 1998. Statistics and Experimental Design for Toxicologists, 3rd Ed. Boca Raton, FL: CRC.

Gad, S.C., and C.P. Chengelis. 1998. Acute Toxicology Testing, 2nd Ed. San Diego, CA: Academic Press.

Garcia, M.M., P.A. Casaco, V.L. Arruzazabala, A.R. Gonzalez, and R.A. de la Vega. 1988. Role of chemical mediators in bronchoconstriction induced by kerosene. Allergol. Immunopathol. 16(6):421-423.

Gaworski, C.L. J.D. MacEwen, E.H. Vernot, R.H. Bruner, and M.J. Cowan, Jr. 1984. Comparison of the subchronic inhalation toxicity of petroleum and oil shale JP-5 jet fuels. Pp. 33-48 in Advances in Modern Environmental Toxicology. Vol. 6: Applied Toxicology of Petroleum Hydrocarbons, H.N. MacFarland, C.E. Holdsworth, J.A. MacGregor et al., eds. Princeton, NJ: Princeton Scientific Publishers.

Gaylor, D.W. 1983. The use of safety factors for controlling risk. J. Toxicol. Environ. Health 11(3):329-336.

Gaylor, D.W., and W. Slikker, Jr. 1990. Risk assessment for neurotoxic effects. Neurotoxicology 11(2):211-218.

Gaylor, D.W., D.M. Sheehan, J.F. Young, and D.R. Mattison. 1988. The threshold dose question in teratogenesis [letter]. Teratology 38(4):389-391.

Generoso, W.M., S.K.Stout, and S.W. Huff. 1971. Effects of alkylating chemicals on reproductive capacity of adult female mice. Mutat. Res. 13(2):172-184.

Goodwin, S.R., L.S. Berman, B.B. Tabeling, and S.F. Sundlof. 1988. Kerosene aspiration: Immediate and early pulmonary and cardiovascular effects. Vet. Hum. Toxicol. 30(6):521-524.

Gordis, L. 1996. Epidemiology. Philadelphia: W.B. Saunders.

Gulati, D.K., E. Hope, J. Teague, and R.E. Chapin. 1991. Reproductive toxicity assessment by continuous breeding in Sprague-Dawley rats: A comparison of two study designs. Fundam. Appl. Toxicol. 17(2):270-279.

Gulson, B.L., C.W. Jameson, K.R. Mahaffey, K.J. Mizon, M.J. Korsch, and G. Vimpani. 1997. Pregnancy increases mobilization of lead from maternal skeleton. J. Lab. Clin. Med. 130(1):51-62.

Haney, A.F., S.F. Hughes, and C.L. Hughes, Jr. 1984. Screening of potential reproductive toxicants by use of porcine granulosa cell cultures. Toxicology 30(3):227- 241.

Hansen, D.K., J.B. LaBorde, K.S. Wall, R.R. Holson, and J.F. Young. 1999. Pharmacokinetic considerations of dexamethasone-induced developmental toxicity in rats. Toxicol. Sci. 48:230-239.

Hardy, C.J., I.J. Sharman, and G.C. Clark. 1991. Assessment of Cardiac Sensitisation Potential in Dogs. Rep. No CTL/C/2521. Huntingdon Research Centre, Huntingdon, Cambridgeshire, U.K.

Harris, C. 1997. In vitro methods for the study of mechanisms of developmental toxicity. Pp. 465-509 in Handbook of Developmental Toxicology, R.D. Hood, ed. Boca Raton, FL: CRC Press.

Harris, M.W., R.E. Chapin, A.C. Lockhart, and M.P. Jokinen. 1992. Assessment of a short-term reproductive and developmental toxicity screen. Fundam. Appl. Toxicol. 19(2):186-196.

Harris, D.T., D. Sakiestewa, R.F. Robledo, and M. Witten. 1997. Immunotoxicological effects of JP-8 jet fuel exposure. Toxicol. Ind. Health 13(1):43-55.

Hathaway, G.J., J.P. Hughes, and N.P. Proctor. 1996. Proctor and Hughes' Chemical Hazards of the Workplace, 4th Ed. New York: John Wiley & Sons.

Hays, A.M., B.J. Tollinger, J.P. Tinajero, R.F. Robledo, R.C. Lantz, and M.L. Witten. 1994. Changes in lung permeability after chronic exposure to JP-8 jet fuel [abstract]. FASEB J 8(4-5):A122.

Heindel, J.J. 1999. Oocyte quantitation and ovarian histology. Pp. 57-74 in An Evaluation and Interpretation of Reproductive Endpoints for Human Health Risk Assessment, G. Daston and C. Kimmel, eds. Washington, DC: ILSI.

Heindel, J.J., C.J. Price, and B.A. Schwetz. 1994. The developmental toxicity of boric acid in mice, rats, and rabbits. Environ. Health Perspect. 102 (Suppl. 7):107-112.

Hemminki, K., and P. Vineis. 1985. Extrapolation of the evidence on teratogenicity of chemicals between humans and experimental animals: Chemicals other than drugs. Teratog. Carcinog. Mutagen. 5(4):251-318.

Hennekens, C., and J.H. Buring. 1987. Need for large sample sizes in randomized trials. Pediatrics 79(4):569-571.

Hennekens, C.H., J.E. Buring, and S.L. Mayrent. 1987. Epidemiology in Medicine. Boston: Little, Brown & Co.

Hext, P.M. 1989. 90-Day Inhalation Toxicity Study in the Rat. ICI Rep. No. CTL/P/ 2466. Central Toxicology Laboratory, Imperial Chemical Industries, Alderley Park, Macclesfield, Cheshire, U.K.

Hext, P.M., and R.J. Parr-Dobrzanski. 1993. HFC-134a: A 2-Year Inhalation Toxicity Study in the Rat. ICI Rep. No. CTL/P/3841. Central Toxicology Laboratory, Imperial Chemical Industries, Alderley Park, Macclesfield, Cheshire, U.K.

Hill, A.B. 1965. The environment and disease: Association or causation? Proc. R. Soc. Med. 58:295-300.

Hodge, M.C.E., M. Kilmartin, R.A. Riley, T.M. Weight, and J. Wilson. 1979a. Arcton 134a: Teratogenicity Study in the Rat. ICI Rep. No. CTL/ P/417. Central Toxicology Laboratory, Imperial Chemical Industries, Alderley Park, Macclesfield, Cheshire, U.K.

Hodge, M.C.E., D. Anderson, I.P. Bannet, and T.M. Whight. 1979b. Arcton-134a: Dominant Lethal Study in the Mouse. ICI Rep. No. CTL/ P/437.

Central Toxicology Laboratory, Imperial Chemical Industries, Alderley Park, Macclesfield, Cheshire, U.K.

Holmes, L.B. 1997. Impact of the detection and prevention of developmental abnormalities in human studies. Reprod. Toxicol. 11(2/3):267-269.

Honeycutt, A., L. Dunlap, H. Chen, and G. al Homsi. 1999. The Cost of Developmental Disabilities. Final report for the Centers for Disease Control and Prevention, Task Order No. 0621-09. RTI No. 6900-009. Research Triangle Institute, Research Triangle Park, NC.

Houston, D.E. 1984. Evidence for the risk of pelvic endometriosis by age, race and socioeconomic status. Epidemiol. Rev. 6:167-191.

HSDB (Hazardous Substances Data Bank). 1998. 1,1,1,2-Tetraflouroethane. CAS No. 811-97-2 in TOXNET (Toxicology Data Network) sponsored by the National Library of Medicine, Toxicology and Environmental Health Information System. [Online]. Available: http://toxnet.nlm.nih.gov/cgi-bin/sis/search [9/28/1998, last updated 8/6/1998].

Hughes, C.L. 1988. Effects of phytoestrogens on GnRH-induced luteinizing hormone secretion in ovariectomized rats. Reprod. Toxicol. 1(3):179-181.

Inouye, M. 1976. Differential staining of cartilage and bone in fetal mouse skeletons by alcian blue and alizarin red S. Congenital Abnormalities 16:171-173.

IOM (Institute of Medicine). 1999. Strategies to Protect the Health of Deployed U.S. Forces: Medical Surveillance, Record Keeping and Risk Reduction. L.M. Joellenbeck, P.K. Russell, and S.B. Guze, eds. Washington, DC: National Academy Press.

IPCS (International Programme on Chemical Safety). 1994. Report of IPCS Workshop on the Harmonization of Risk Assessment for Reproductive and Developmental Toxicity. Oct. 17-21, 1994. International Programme on Chemical Safety, World Health Organization, Geneva, Switzerland.

IPCS (International Programme on Chemical Safety). 1995. IPCS/OECD Workshop on the Harmonization of Risk Assessment for Reproductive and Development Toxicity. BIBRA International, Carshalton, U.K, Oct. 18-21, 1994, IPCS/95.25. International Programme on Chemical Safety, World Health Organization, Geneva, Switzerland.

IRIS (Integrated Risk Information System). 1998. 1,1,1,2-Tetraflouroethane. CAS No. 811-97-2 in TOXNET(Toxicology Data Network) sponsored by the National Library of Medicine, Toxicology and Environmental Health Information System. [Online]. Available: http://toxnet.nlm.nih.gov/cgi-bin/sis/search [9/25/1998, last updated 8/6/1998].

Jee, S.H., J.D. Wang, C.C. Sun, and Y.F. Chao. 1985. Prevalence of probable kerosene dermatoses among ball-bearing factory workers. Scand. J. Work Environ. Health 12(1):61-65.

Jekel, J.F., J.G. Elmore, and D.L. Katz. 1996. Epidemiology, Biostatistics and Preventive Medicine. Philadelphia: W.B. Saunders & Co.

Joffe, M. 1985. Biases in research on reproduction and women's work. Int. J. Epidemiol. 14(1):118-123.

Johlin, F.C., C.S. Fortman, D.D. Nghiem, and T.R. Tephly. 1987. Studies on the role of folic acid and folate-dependent enzymes in human methanol poisoning. Mol. Pharmacol. 31(5):557-561.

Kennedy, G.L. 1979. Subacute Inhalation Toxicity of Tetrafluoroethane (FC 134a). Haskell Laboratory. Report no. 228-79. Du Pont de Nemours and Company, Newark, DE.

Kimmel, C.A., and D.W. Gaylor. 1988. Issues in qualitative and quantitative risk analysis for developmental toxicology. Risk Anal. 8(1):15-20.

Kimmel, C.A., and E.Z. Francis. 1990. Proceedings of the workshop on the acceptability and interpretation of dermal developmental toxicity studies. Fundam. Appl. Toxicol. 14(2):386-398.

Kimmel, C.A., D.C. Rees, and E.Z. Francis, eds. 1990. Qualitative and Quantitative Comparability of Human and Animal Developmental Neurotoxicity. Special Issue. Neurotoxicology and Teratology 12(3):173-285.

Kimmel, C.A., and J.F. Young. 1983. Correlating pharmacokinetics and teratogenic endpoints. Fundam. Appl. Toxicol. 3(4):250-255.

Kimmel, G.L. 1990. In vitro assays in developmental toxicology: Their potential application in risk assessment. Pp.163-173 in In Vitro Methods in Developmental Toxicology: Use in Defining Mechanisms and Risk Parameters, G.L. Kimmel, and D. M. Kochhar, eds. Boca Raton, FL: CRC Press.

Kimmel, G.L., and D. M. Kochhar. 1990. In Vitro Methods in Developmental Toxicology: Use in Defining Mechanisms and Risk Parameters. Boca Raton, FL: CRC Press.

Kimmel, C.A., J.F. Holson, C.J. Hogue, and G. Carlo. 1984. Reliability of Experimental Studies for Predicting Hazards to Human Development. NCTR Technical Report for Experiment No. 6015. National Center for Toxicological Research, Jefferson, AR.

Kimmel, G.L., K. Smith, D.M. Kochhar, and R.M. Pratt. 1982. Overview of In Vitro Teratogenicity Testing: Aspects of validation and application to screening. Teratog. Carcinog. Mutagen. 2(3-4):221-229.

Kinkead, E.R., S.A. Salins, and R.E. Wolfe. 1984. Acute irritation and sensitization potential of JP-8 fuel. J. Am. Coll. Toxicol. 11(6):700.

Kistler, A. 1987. Limb bud cell cultures for estimating the teratogenic potential of compounds. Validation of the test system with retinoids. Arch. Toxicol. 60(6): 403-414.

Knave, B, B.A. Olson, S. Elofsson, F. Gamberale, A. Isaksson, P. Mindus, H.E. Persson, G. Struwe, A. Wennberg, and P. Westerholm. 1978. Long-term exposure to jet fuel: II. A cross-sectional epidemiologic investigation on occupationally exposed industrial workers with special reference to the nervous system. Scand. J. Work Environ. Health 4(1):19-45.

Kodell, R.L., J.J. Chen, and D.W. Gaylor. 1995. Neurotoxicity modeling for risk assessment. Regul. Toxicol. Pharmacol. 22(1):24-29.

Lamb, J.C., IV. 1985. Reproductive toxicity testing: Evaluating and developing new testing systems. J. Am. Coll. Toxicol. 4(2):163-171.

Lamb, J.C., IV. 1988. Fundamentals of male reproductive toxicity testing. Pp. 137-153 in Physiology and Toxicology of Male Reproduction, J.C. Lamb, IV, and P.M.D. Foster, eds. New York: Academic Press.

Lamb, J.C., IV. 1989. Design and use of multigeneration breeding studies for identification of reproductive toxicants. Pp. 131-155 in Toxicology of the Male and Female Reproductive Systems, P.K. Working, ed. New York: Hemisphere Publishing.

Last, J.M. 1995. A Dictionary of Epidemiology, 3rd Ed. New York: Oxford University Press.

Lemasters, G.K., J.E. Lockey, D.M. Olsen, S.G. Selevan, M.W. Tabor, G.K. Livingston, and G.R. New. 1999a. Comparison of internal dose measures of solvents in breath, blood and urine and genotoxic changes in aircraft maintenance personnel. Drug Chem. Toxicol. 22(1):181-200.

Lemasters, G.K., D.M. Olsen, J.H. Yiin, J.E. Lockey, R. Shukla, S.G. Selevan, S.M. Schrader, G.P. Toth, D.P. Evenson, and G.B. Huszar. 1999b. Male reproductive effects of solvent and fuel exposure during aircraft maintenance. Reprod. Toxicol. 13(3):155-166.

Lewis, S.C., J.R. Lynch, and A.I. Nikiforov. 1990. A new approach to deriving community exposure guidelines from "no-observed-adverse-effect levels." Regul. Toxicol. Pharmacol. 11(3):314-330.

Lilienfeld, A.M. 1959. On the methodology of investigations of etiologic factors in chronic diseases—Some comments. J. Chronic Dis. 10(1):41-46.

Litton Bionetics. 1976. Mutagenicity Evaluation of Genetron-134a. Final Report. LBI Project No. 2683. Prepared for Allied Chemical Corp. by Litton Bionetics, Kensington, MD.

Longstaff, E., M. Robinson, C. Bradbrook, J.A. Styles, and I.F.H. Purchase. 1984. Genotoxicity and carcinogenicity of fluorocarbons: Assessment by short-term in vitro tests and chronic exposure in rats. Toxicol. Appl. Pharmacol. 72(1):15-31.

Lu, M.H., and R.E. Staples. 1981. 1,1,1,2-Tetrafluoroethane (FC-134a): Embryo-Fetal Toxicity and Teratogenicity Study by Inhalation in the Rat. Haskell Laboratory Rep. No. 317-81. Haskell Laboratory for Toxicology and Industrial Medicine, Central Research and Development Department, Newark, DE.

Luecke, R.H., W.D. Wosilait, B.A. Pearce, and J.F. Young. 1994. A physiologically based pharmacokinetic computer model for human pregnancy. Teratology 49(2):90-103.

Mackay, J.M. 1990. HFC-134a: An Evaluation in the In Vitro Cytogenetic Assay in Human Lymphocytes. ICI Rep. No. CTL/P/2977. Central Toxicology Laboratory, Imperial Chemical Industries, Alderley Park, Macclesfield, Cheshire, U.K.

MacMahon, B., and T.F. Pugh. 1970. Epidemiology; Principles and Methods. Boston: Little, Brown.

Mattie, D.R., C.L. Alden, T.K. Newell, C.L. Gaworski, and C.D. Flemming. 1991. A 90-day continuous vapor inhalation toxicity study of JP-8 jet fuel followed by 20 or 21 months of recovery in Fischer 344 rats and C57BL/6 mice. Toxicol. Pathol. 19(2):77-87.

Mattie, D.R., G.B. Marit, C.D. Flemming, and J.R. Cooper. 1995. The effects of JP-8 jet fuel on male Sprague-Dawley rats after a 90-day exposure by oral gavage. Toxicol. Ind. Health 11(4):423-435.

May, P.C., and C.E. Finch. 1988. Aging and responses to toxins in female reproductive functions. Reprod. Toxicol. 1(3):223-228.

McLachlan, J.A., R.R. Newbold, K.S. Korach, J.C. Lamb, IV, and Y. Suzuki. 1981. Transplacental toxicology: prenatal factors influencing postnatal fertility. Pp. 213-232 in Developmental Toxicology, C.A. Kimmel, and J. Buelke-Sam, eds. New York: Raven Press.

Medinsky, M.A., and D.C. Dorman. 1995. Recent developments in methanol toxicity. Toxicol. Lett. 82-83:707-711.

Mehm, W.J., and C.L. Feser. 1984. Biological analysis of progressive toxicity of shale-derived vs. petroleum derived fuels in rats. Pp 491-503 in Synthetic Fossil Fuel Technologies: Results of Health and Environmental Studies, K.E. Cowser, ed. Boston: Butterworth.

Meistrich, M.L. 1986. Critical components of testicular function and sensitivity to disruption. Biol. Reprod. 34(1):17-28.

Mercier, O. 1989. HFA-134a: Test to Determine the Index of Primary Cutaneous Irritation in the Rabbit. Rep. No. 911422. Hazelton Laboratories, France.

Mercier, O. 1990a. HFA-134a: Test to Evaluate the Ocular Irritation in the Rabbit. Rep. No. 912349. Hazelton Laboratories, France.

Mercier, O. 1990b. HFA-134a: Test to Evaluate the Sensitising Potential by Topical Applications in the Guinea Pig. The Epicutaneous Maximisation Test. Rep. No. 001380. Hazelton Laboratories, France.

Miller, M.T., and K. Strömland. 1999. Teratogen update: Thalidomide: A review, with a focus on ocular findings and new potential uses. Teratology 60(5):306-321.

Moller, H., and N.E. Skakkebaek. 1999. Risk of testicular cancer in subfertile men: Case-control study. BMJ 318(7183):559-562.

Moore, J.A., G.P. Daston, E. Faustman, M.S. Golub, W.L. Hart, C. Hughes, Jr., C.A. Kimmel, J.C. Lamb, IV, B.A. Schwetz, and A.R. Scialli. 1995a. An evaluative process for assessing human reproductive and developmental toxicity of agents. Reprod. Toxicol. 9(1):61-95.

Moore, J.A., and IEHR Expert Scientific Committee. 1995b. An assessment of lithium using the IEHR evaluative process for assessing human developmental and reproductive toxicity of agents. Reprod. Toxicol. 9(2):175-210.

Moore, J.A., and an Expert Scientific Committee. 1997. An assessment of boric acid and borax using the IEHR evaluative process for assessing human developmental and reproductive toxicity of agents. Reprod. Toxicol. 11(1):123-160.

Moore, K.L. 1988. The Developing Human, 4th Ed. Philadelphia: W.B. Saunders.

Morrissey, R.E., J.C. Lamb, IV, B.A. Schwetz, J.L. Teague, and R.W. Morris. 1988a. Association of sperm, vaginal cytology, and reproductive organ weight data with results of continuous breeding reproduction studies in Swiss (CD-1) mice. Fundam. Appl. Toxicol. 11(2):359-371.

Morrissey, R.E., B.A. Schwetz, J.C. Lamb, IV, M.D. Ross, J.L. Teague, and R.W. Morris. 1988b. Evaluation of rodent sperm, vaginal cytology, and reproductive organ weight data from National Toxicology Program 13-week studies. Fundam. Appl. Toxicol. 11(2):343-358.

Mostofi, F.K., and E.B. Price. 1973. Tumours of the testis, Leydig cell tumor. Pp 86-99 in Tumours of the Male Genital System. Reprinted 1987. Fascicle 8, Armed Forces Institute of Pathology.

Muller, W., and T. Hoffmann. 1989. CFC-134a Micronucleus Test in Male and Female Mice after Inhalation. Rep. No. 89.0015, Study No. 88.1244. Hoechst Aktiengesellschaft, Frankfurt, Germany.

Mullin, L.S., and R.W. Hartgrove. 1979. Cardiac Sensitization. Rep. No. 42-79. Haskell Laboratory, Wilmington, DE.

Mullin, L.S., C.F. Reinhardt, and R.E. Hemingway. 1979. Cardiac arrhythmias and blood levels associated with inhalation of Halon 1301. Am. Ind. Hyg. Assoc. J. 40(7):653-658.

Muralidhara, M.K. Krishnakumari, H.P. Ramesh, S.K. Mysore, and Association of Food Scientists and Technologists. 1982. Toxicity of some petroleum fractions used in pesticidal emulsions to albino rats.(India). J. Food Sci. Technol. 19(6):260-262.

Narotsky, M.G., J.E. Schmid, J.E. Andrews, and R.J. Kavlock . 1998. Effects of boric acid on axial skeletal development in rats. Biol. Trace Elem. Res. 66(1-3):373- 394.

Nau, H. 1986. Species differences in pharmacokinetics and drug teratogenesis. Environ. Health Perspect. 70:113-129.

Nau, H., and W. J. Scott, Jr., eds. 1987. Pharmacokinetics in Teratogenesis, Vols. 1 and 2. Boca Raton, FL: CRC Press.

Nelson, C.J., and J.F. Holson. 1978. Statistical analysis of teratologic data: Problems and advancements. J. Environ. Pathol. Toxicol. 2(1):187-199.

Nelson, K., and L.B. Holmes. 1989. Malformations due to presumed spontaneous mutations in newborn infants. N. Engl. J. Med. 320(1):19-23.

Newman, L.M., E.M. Johnson, and R.E. Staples. 1993. Assessment of the effectiveness of animal developmental toxicity testing for human safety. Reprod. Toxicol. 7(4):359-390.

NIOSH (National Institute for Occupational Safety and Health). 1999. NIOSH Pocket Guide to Chemical Hazards. National Institute for Occupational Safety and Health. [Online]. Available: http://www.cdc.gov/niosh/npg/npgd0438.html.

NIOSH (National Institute for Occupational Safety and Health). 2000. NIOSH Pocket Guide to Chemical Hazards (NPG). National Institute for Occupational Safety and Health, Washington, DC. [Online]. Available: wysiwyg://17/http://www.cdc.gov/niosh/npg/npg.html.

Nisbet, I.C.T., and N.J. Karch. 1983. Chemical Hazards to Human Reproduction. Noyes Data Corp., Park Ridge, IL.

Noa, M., and J. Illnait. 1987a. Changes in the aorta of guinea pigs exposed to kerosene. Acta Morphol. Hung. 35(1-2):59-70.

Noa, M., and J. Illnait. 1987b. Induction of aortic plaques in guinea pigs by exposure to kerosene. Arch. Environ. Health 42(1):31-36.

Nouri, L.A., D.O. Sordelli, M.C. Cerquetti, J.M. Saavedra, A.M. Hooke, and J.A. Bellanti. 1983. Pulmonary clearance of *Staphylococcus aureus* and plasma angiotensin-converting enzyme activity in hydrocarbon pneumonitis. Pediatr. Res. 17(8):657-661.

NRC (National Research Council). 1989. Biologic Markers in Reproductive Toxicology. Washington, DC: National Academy Press.

NTP/NIH (National Toxicology Program/National Institutes of Health). 1986. Toxicology and Carcinogenesis Studies of Marine Diesel Fuel and JP-5 Navy Fuel in B6C3F1 Mice (dermal study). National Toxicology Program Technical Report Series No 310. NIH Publ. No. 86-2566. National Toxicology Program/National Institutes of Health, Research Triangle Park, NC.

OECD (Organization for Economic Cooperation and Development). 1983. OECD Guidelines for Testing of Chemicals. One-Generation Reproduction Toxicity Study. Adopted Guideline. TG No. 415. Organization for Economic Cooperation and Development, Paris, France.

OECD (Organization for Economic Cooperation and Development). 1984. OECD Guidelines for Testing of Chemicals. Genetic Toxicology: Rodent

Dominant Lethal Test. TG No. 478. Organization for Economic Cooperation and Development, Paris France.

OECD (Organization for Economic Cooperation and Development). 1987. OECD Guidelines for Testing of Chemicals. Organization for Economic Cooperation and Development, Paris, France.

OECD (Organization for Economic Cooperation and Development). 1995. OECD Guidelines for Testing of Chemicals. Reproduction/Developmental Toxicity Screening Test. TG No. 421. Organization for Economic Cooperation and Development, Paris, France.

OECD (Organization for Economic Cooperation and Development). 1996. OECD Guidelines for Testing of Chemicals. Combined Repeated Dose Toxicity Study with the Reproduction/ Developmental Toxicity Screening Test. TG No. 422. Organization for Economic Cooperation and Development, Paris, France.

OECD (Organization for Economic Cooperation and Development). 2000a. OECD Guidelines for Testing of Chemicals. Prenatal Developmental Toxicity Study. Approved Draft. TG No. 414. ENV/EPOC(2000)9. Organization for Economic Cooperation and Development, Paris, France. June. [Online]. Available: http://www.oecd.org//ehs/test/health. htm#DRAFT GUIDELINES

OECD (Organization for Economic Cooperation and Development). 2000b. OECD Guidelines for Testing of Chemicals. Two-Generation Reproduction Toxicity Study. Approved Draft. TG No. 416. Organization for Economic Cooperation and Development, Paris, France. June. [Online]. Available: http://www.oecd.org//ehs/test/health.htm#DRAFT GUIDELINES

O'Flaherty, E.J., W. Scott, C. Schreiner, and R.P. Beliles. 1992. A physiologically based kinetic model of rat and mouse gestation: Disposition of a weak acid. Toxicol. Appl. Pharmacol. 112(2):245-256.

Ohta, H., I. Sugimoto, A. Masuda, S. Komukai, Y. Suda, K. Makita, K. Takamatsu, F. Horiguchi, and S. Nozawa. 1996. Decreased bone mineral density associated with early menopause progresses for at least ten years: Cross-sectional comparisons between early and normal menopausal women. Bone 18(3):227-231.

Olive, D.L., and L.B. Schwartz. 1993. Endometriosis. N. Engl. J. Med. 328(24):1759- 1769.

Olshan, A.F., D.R. Mattison, and T.S. Zwanenburg. 1994. International Commission for Protection Against Environmental Mutagens and Carcinogens. Cyclosporine A: Review of genotoxicity and potential for adverse human reproductive and developmental effects. Report of a Working Group on the genotoxicity of cyclosporine A, August 18, 1993. Mutat. Res. 317(2):163-173.

Olson, M.J., C.A. Reidy, J.T. Johnson, and T.C. Pederson. 1990. Oxidative defluorination of 1,1,1,2-tetrafluoroethane by rat liver microsomes. Drug Metab. Dispos. 18:992-998.

Palmer, A.K. 1986. Experimental teratology, introduction. Teratology 34(3): 405.

Palmer, A.K., and B.C. Ulbrich. 1997. The cult of culling. Fundam. Appl. Toxicol. 38(1):7-22.

Parker, G.A., V. Bogo, and R.W. Young. 1981. Acute toxicity of conventional versus shale-derived JP5 jet fuel: Light microscopic, hematologic, and serum chemistry studies. Toxicol. Appl. Pharmacol. 57(3):302-317.

Parvinen, M. 1982. Regulation of the seminiferous epithelium. Endocr. Rev. 3(4):404-417.

Pederson, T., and H. Peters. 1968. Proposal for a classification of oocytes and follicles in the mouse ovary. J. Reprod. Fertil. 17:555-557.

Pfaff, J., G. Parliman, K. Parton, R. Lantz, H. Chen, M. Hays, and M. Witten. 1993. Pathologic changes after JP-8 fuel inhalation in Fischer 344 rats [abstract]. FASEB J. 7(3-4):A408.

Pleil, J.D., L.B. Smith, and S.D. Zelnick. 2000. Personal exposure to JP-8 jet fuel vapors and exhaust at air force bases. Environ. Health. Perspect. 108(3):183-92.

Porter, H.O. 1990. Aviators intoxicated by inhalation of JP-5 fuel vapors. Aviat. Space Environ. Med. 61(7):654-656.

Price, C.J., P. L. Strong, M.C. Marr, C.B. Myers, and F.J. Murray. 1996. Developmental toxicity NOAEL and postnatal recovery in rats fed boric acid during gestation. Fundam. Appl. Toxicol. 32(2):179-193.

Puhala, E., G. Lemasters, L. Smith, G. Talaska, S. Simpson, J. Joyce, K. Trinh, and J. Lu. 1997. Jet fuel exposure in the United States Air Force. Appl. Occup. Environ. Hyg. 12(9):606-610.

Renwick, A.G. 1991. Safety factors and establishment of acceptable daily intakes. Food Addit. Contam. 8(2):135-149.

Renwick, A.G. 1993. Data-derived safety factors for the evaluation of food additives and environmental contaminants. Food Addit. Contam. 10(3): 275-305.

Renwick, A.G. 1998. Toxicokinetics in infants and children in relation to the ADI and TDI. Food Addit. Contamin. 15(Suppl.):17-35.

Renwick, A.G., and N.R. Lazarus. 1998. Human variability and noncancer risk assessment: An analysis of the default uncertainty factor. Regul. Toxicol. Pharmacol. 27(1 Part 1):3-20.

Reidy, C.A., J.T. Johnson, and M.J. Olson. 1990. Metabolism in vitro of fluorocarbon R-134a. [Abstract 1295]. Toxicologist 10(1):324.

Riley, R.A., I.P. Bennett, I.S. Chart, C.W. Gore, M. Robinson, and T.M. Weight.

1979. Arcton-134a: Subacute Toxicity to the Rat by Inhalation. ICI Rep. No. CTL/P/463. Central Toxicology Laboratory, Imperial Chemical Industries, Alderley Park, Macclesfield, Cheshire, U.K.

Rissolo, S.B., and J.A. Zapp. 1967. Acute Inhalation Toxicity. Rep. No. 190-67. Haskell Laboratory, Wilmington, DE.

Riviere, J.E., J.D. Brooks, N.A. Monteiro-Riviere, K. Budsaba, and C.E. Smith. 1999. Dermal absorption and distribution of topically dosed jet fuels jet-A, JP-8, and JP-8(100). Toxicol. Appl. Pharmacol. 160(1):60-75.

Rogers, J.M., M.L. Mole, N. Chernoff, B.D. Barbee, C.I. Turner, T.R. Logsdon, and R.J. Kavlock. 1993. The developmental toxicity of inhaled methanol in the CD-1 mouse, with quantitative dose-response modeling for estimation of benchmark doses. Teratology 47(3):175-188.

Rothman, K.J., and S. Greenland. 1998. Pp 160-161 in Modern Epidemiology. 2nd Ed. Philadelphia, PA: Lippincott-Raven Publishers.

Safe, S. 1993. Development of bioassays and approaches for the risk assessment of 2,3,7,8-tetrachlorodibenzo-p-dioxin and related compounds. Environ. Health Perspect. 101(Suppl. 3):317-325.

Salewski, V. 1964. Faerbermethode zum makroskopischen nachweis von implantations stellen am uterus der ratte. Naunyn-Schmeidebergs Arch. Pharmakol. Exp. Pathol. 247:367.

Salmon, A.G., J.A. Nash, M.F.S. Oliver, and A. Reeve. 1980. Arcton-134a: Excretion, Tissue Distribution, and Metabolism in the Rat. ICI Rep. No. CTL/P/513. Central Toxicology Laboratory, Imperial Chemical Industries, Alderley Park, Macclesfield, Cheshire, U.K.

Sartwell, P.E. 1960. On the methodology of investigations of etiologic factors in chronic diseases — Further comments. J. Chronic Dis. 11:61-63.

Schardein, J.L. 2000. Chemically-induced Birth Defects. 3rd Ed. New York: Marcel Dekker.

Schardein, J.L. 1998. Animal/human concordance. Pp. 687-708 in Handbook of Developmental Neurotoxicology, W. Slikker Jr., and L.W. Chang, eds. San Diego, CA: Academic Press.

Schultz, T.W., H. Witschi, L.H. Smith, W.M. Haschek, J.M. Holland, J.L. Epler, R.J. Fry, T.K. Rao, F.W. Larimer, and J.N. Dumont. 1981. Health Effects Research in Oil Shale Development. Technical Report No. ORNL/TM-8034. Oak Ridge National Laboratory, Oak Ridge, TN.

Schwetz, B.A., R.E. Morrissey, F. Welsch, and R.A. Kavlock. 1991. In vitro teratology. Environ. Health Perspect. 94:265-268.

Scialli, A.R., and A. Leone. 1998. Variability in human response to reproductive and developmental toxicity. Pp. 87-137 in Human Variability in Response to Chemical Exposures: Measures, Modeling and Risk Assessment, D.A. Neumann, and C.A. Kimmel, eds. Washington, DC: ILSI.

Scialli, A.R., A. Leone, and G.K. Boyle Padgett. 1995. Reproductive Effects of Chemical, Physical, and Biologic Agents: Reprotex. Baltimore, MD: Johns Hopkins University Press.

Selevan, S.G., C.A. Kimmel, and P. Mendola. 2000. Identifying Critical Windows of Exposure for Children's Health. Monograph based on papers developed from the Workshop to Identify Critical Windows for Exposure for Children's Health held 14-15 September 1999 in Richmond, VA. Environ. Health Perspect. 108(Suppl. 3):451-597.

Shepard, T.H. 1998. Catalog of Teratogenic Agents. Baltimore, MD: Johns Hopkins University Press.

Shuey, D.L., C. Lau, T.R. Logsdon, R.M. Zucker, K.H. Elstein, M.G. Narotsky, R.W. Setzer, R.J. Kavlock, and J.M. Rogers. 1994. Biologically based dose- response modeling in developmental toxicology: Biochemical and cellular sequelae of 5-fluorouracil exposure in the developing rat. Toxicol. Appl. Pharmacol. 126(1):129-144.

Shulman, M., and M.S. Sadove. 1967. 1,1,1,2-Tetrafluoroethane: An inhalation anaesthetic agent of intermediate potency. Anesth. Analg. 46(5):629-633.

Silber, L.S., and G.L. Kennedy. 1979a. Acute Inhalation Toxicity of Tetrafluoroethane. Rep. No. 422-79. Haskell Laboratory, Wilmington, DE.

Silber, L.S., and G.L. Kennedy. 1979b. Subacute Inhalation Toxicity of Tetrafluoroethane (HFC 134a). Rep. No. 228-79. Haskell Laboratory, Wilmington, DE.

Silbergeld, E.K. 1991. Lead in bone: Implications for toxicology during pregnancy and lactation. Environ. Health Perspect. 91:63-70.

Silbergeld, E.K., J. Schwartz, and K. Mahaffey. 1988. Lead and osteoporosis: Mobilization of lead from bone in postmenopausal women. Environ. Res. 47(1):79-94.

Spielmann, H., G. Scholtz, A. Seiler, I. Pohl, S. Bremer, N. Brown, and A. Piersma. 1998. Ergebnisse der ersten Phase des ECVAM-Projektes zur Prävalidierung und Validierung von drei in vitro embryotoxizitätstests. ALTEX 15:3-8.

Starek, A., and M. Vojtisek. 1986. Effects of kerosine hydrocarbons on tissue metabolism in rats. Pol. J. Pharmacol. Pharm. 38(5-6):461-470.

Stephen, E.H. 1996. Projections of impaired fecundity among women in the United States: 1995 to 2020. Fertil Steril. 66(2):205-209.

Stephen, E.H., and A. Chondra. 1998. Updated projections of infertility in the United States: 1995-2025. Fertil Steril. 70(1):30-34.

Strömland, K, V. Nordin, M. Miller, B. Akerstrom, and C. Gillberg. 1994. Autism in thalidomide embryopathy: A population study. Dev. Med. Child. Neurol. 36(4):351-356.

Struwe, G., B. Knave, and P. Mindus. 1983. Neuropsychiatric symptoms in workers occupationally exposed to jet fuel-A combined epidemiological and casuistic study. Acta Psychiatr. Scand. (Suppl.):303:55-67.

Sullivan, F.M., S.E. Smith, and P.R. McElhatton. 1987. Interpretation of animal experiments as illustrated by studies on caffeine. Pp. 123-130 in Pharmacokinetics in Teratogenesis, Vol 1. Interspecies Comparison and Maternal / embryonic-fetal Drug Transfer, H. Nau, and W.J. Scott, eds. Boca Raton, FL: CRC Press.

Surgeon General's Advisory Committee on Smoking and Health. 1964. Smoking and Health. DHEW pub. no. PHS 1103. Rockville, MD: U.S. Public Health Service.

Susser, M. 1973. Causal Thinking in the Health Sciences. New York: Oxford University Press.

Teaff, N.L., R.T. Savoy-Moore, M.G. Subramanian, and K.M. Ataya. 1990. Vinblastine reduces progesterone and prostaglandin E production by rat granulosa cells in vitro. Reprod. Toxicol. 4(3):209-213.

Teng, C.T., M.P. Walker MP, S.N. Bhattacharyya, D.G. Klapper, R.P. DiAugustine, and J.A. McLachlan. 1986. Purification and properties of an oestrogen-stimulated mouse uterine glycoprotein (approx. 70 kDa). Biochem. J. 240(2): 413-422.

Tephly, T.R., and K.E. McMartin. 1984. Methanol metabolism and toxicity. Pp. 111-140 in Aspartame: Physiology and Biochemistry, L.D. Stegink, and L.J. Filer, Jr., eds. New York: Marcel Dekker.

Terry, K.K., B.A. Elswick, D.B. Stedman, and F. Welsch. 1994. Developmental phase alters dosimetry-teratogenicity relationship for 2-methoxyethanol in CD-1 mice. Teratology 49:218-227.

Thomas, J.A. 1981. Reproduction hazards and environmental chemicals: a review. Toxic Subst. J. 2(4):318-348.

Toppari, J., P.C. Bishop, J.W. Parker, N. Ahmad, W. Girgis, and G.S. diZerega. 1990. Cytotoxic effects of cyclophosphamide in the mouse seminiferous epithelium: DNA flow cytometric and morphometric analysis. Fundam. Appl. Toxicol. 15(1):44-52.

Trochimowicz, H.J., A. Azar, J.B. Terrill, and L.S. Mullin. 1974. Blood levels of fluorocarbon related to cardiac sensitization: Part II. Am. Ind. Hyg. Assoc. J. 35(10):632-639.

Trueman, R.W. 1990. HFC-134a: Assessment for the Induction of Unscheduled DNA Synthesis in Rat Hepatocytes In Vivo. ICI Rep. No CTL/P/2550. Central Toxicology Laboratory, Imperial Chemical Industries, Alderley Park, Macclesfield, Cheshire, U.K.

Tukey, J.W., J.L. Ciminera, and J.F. Heyse. 1985. Testing the statistical certainty of a response to increasing doses of a drug. Biometrics 41(1): 295-301.

Tzimas, G., R. Thiel, I. Chahoud, and H. Nau. 1997. The area under the concentration-time curve of all-trans-retinoic acid is the most suitable pharmacokinetic correlate to the embryotoxicity of this retinoid in the rat. Toxicol. Appl. Pharmacol. 143:436–444.

Uphouse, L. 1985. Effects of chlordecone on neuroendocrine function of female rats. Neurotoxicology 6(1):191-210.

Uphouse, L., and J. Williams. 1989. Sexual behavior of intact female rats after treatment with o,p'-DDT or p,p'-DDT. Reprod. Toxicol. 3(1):33–41.

Upreti, R.K., M. Das, and R. Shanker. 1989. Dermal exposure to kerosene. Vet. Hum. Toxicol. 31(1):16-20.

USAF (United States Air Force). 1978a. Mutagen and Oncogen Study on JP-8. Aerospace Medical Research Laboratory, Aerospace Medical Division, Air Force Systems Command, Wright-Patterson Air Force Base, OH. NTIS Publ. No. AD-A064-948/3.

USAF (United States Air Force). 1978b. Toxic Hazards Research Unit Annual Technical Report: 1978. Aerospace Medical Research Laboratory, Aerospace Medical Division, Air Force Systems Command, Wright-Patterson Air Force Base, OH. NTIS Publ. No. AD-A062-138.

USAF (United States Air Force). 1985. Evaluation of the 90-day Inhalation Toxicity of Petroleum and Oil Shale JP-5 Jet Fuel. Aerospace Medical Research Laboratory, Aerospace Medical Division, Air Force Systems Command, Wright-Patterson Air Force Base, OH. NTIS Publ. No. AD-A156-815.

USAF (United States Air Force). 1991. Supercritical Fluid Fractionation of JP-8. Final report for period August 1990-June 1991. Aero Propulsion and Power Directorate, Wright Research Development Center, Air Force Systems Command, Wright-Patterson Air Force Base, OH. NTIS Publ. No. AD-A247-835.

USAF (United States Air Force). 1994. The Chronic Effects of JP-8 Jet Fuel Exposure on the Lungs. Life and Environmental Sciences Directorate, U.S. Air Force Office of Scientific Research, Washington, DC. NTIS Publ. No. AD-A280-982.

U.S. Congress. 1988. Infertility, Medical and Social Choices. OTA-BA-358. Office of Technology Assessment. Washington, DC: U.S. Government Printing Office.

Vinegar, A., G.W. Jepson, R.S. Cook, J.D. McCafferty, and M.C. Caracci. 1997. Human Inhalation of Halon 1301, HFC-134a and HFC-227ea for Collection of Pharmacokinetic Data. U.S. Air Force Armstrong Laboratory. AL/OE-TR-1997-0116. Occupational and Environmental Health Directorate, Toxicology Division, Wright-Patterson Air Force Base, OH.

Weed, D.L. 1995. Causal and preventive inference. Pp 285-302 in Cancer Prevention and Control, P. Greenwald, B.S. Kramer, and D.L. Weed, eds. New York: Marcel Dekker.

Weed, D.L. 1997. Underdetermination and incommensurability in contemporary epidemiology. Kennedy Inst. Ethics J. 7(2):107-127.

Weinberg, C.R., D.D. Baird, and A.J. Wilcox. 1994. Sources of bias in studies of time to pregnancy. Stat. Med. 13(5-7):671-681.

Weller, E., N. Long, A. Smith, P. Williams, S. Ravi, J. Gill, R. Henessey, W. Skornik, J. Brain, C. Kimmel, G. Kimmel, L. Holmes, and L. Ryan. 1999. Dose-rate effects of ethylene oxide exposure on developmental toxicity. Toxicol. Sci. 50(2):259-270.

Whitaker, J., and K.D. Dix. 1979. Double staining for rat foetus skeletons in teratological studies. Lab. Animals 13:309-310.

Wickramaratne, G.A. 1989a. HFC 134a: Embryotoxicity Inhalation Study in the Rabbit. ICI Rep. No. CTL/P/2380. Central Toxicology Laboratory, Alderley Park Macclesfield, Cheshire, U.K.

Wickramaratne, G.A. 1989b. HFC 134a: Teratogenicity Inhalation Study in the Rabbit. ICI Rep. No. CTL/P/2504. Central Toxicology Laboratory, Alderley Park, Macclesfield, Cheshire, U.K.

Williams, D.A. 1971. A test for differences between treatment means when several dose levels are compared with a zero dose control. Biometrics 27(1):103-117.

Williams, D.A. 1972. The comparison of several dose levels with a zero dose control. Biometrics 28(2):519-531.

Wilson, J.G. 1973. Environment and Birth Defects. New York: Academic Press.

Wilson, J.G., W.J. Scott, E.J. Ritter, and R. Fradkin. 1975. Comparative distribution and embryotoxicity of hydroxyurea in pregnant rats and rhesus monkeys. Teratology 11(2):169-178.

Wilson, J.G., E.J. Ritter, W.J. Scott, and R. Fradkin. 1977. Comparative distribution and embryotoxicity of acetylsalicylic acid in pregnant rats and rhesus monkeys. Toxicol. Appl. Pharmacol. 41(1):67-78.

Wise, L.D., S.L. Beck, D. Beltrame, B.K. Beyer, I. Chahoud, R.L. Clark, R. Clark, A.M. Druga, M.H. Feuston, P. Guittin, S.M. Henwood, C.A. Kimmel, P. Lindstrom, A.K. Palmer, J.A. Petrere, H.M. Solomon, M. Yasuda, and R.G. York. 1997. Terminology of developmental abnormalities in common laboratory mammals (version 1). Teratology. 55(4):249-292.

Working, P.K. 1988. Male reproductive toxicity: Comparison of the human to animal models. Environ. Health 77:37-44.

Yacobi, A., J.P. Skelly, V.P. Shah, and L.Z. Benet. 1993. Integration of Pharmacokinetics, Pharmacodynamics, and Toxicokinetics in Rational Drug Development. New York: Plenum Press.

Young, J.F., and J.F. Holson. 1978. Utility of pharmacokinetics in designing toxicological protocols and improving interspecies extrapolation. J. Environ. Pathol. Toxicol. 2(1):169-186.

Zenick, H., and E.D. Clegg. 1989. Assessment of male reproductive toxicity: A risk assessment approach. Pp. 279-309 in Principles and Methods of Toxicology, 2nd Ed., A.W. Hayes, ed. New York: Raven Press.

Zill, N., and C.A. Schoenborn. 1990. Developmental, learning, and emotional problems. Health of our nation's children, United States, 1988. Adv. Data (190):1-18.

Zucker, A.R., S. Berger, and L.D. Wood. 1986. Management of kerosene-induced pulmonary injury. Crit. Care Med. 14(4):303-304.

Appendixes

Appendix A

Application of the Recommended Evaluative Process to Specific Chemicals

This appendix demonstrates the evaluative process described in Chapters 2 and 3. To do that, the Subcommittee on Reproductive and Developmental Toxicology evaluated two agents of interest to the Navy, jet fuel JP-8 and 1,1,1,2-tetrafluoroethane (HFC-134a). The Navy is concerned about the health effects, including reproductive and developmental effects, of exposure to these agents. JP-8 was selected because it is a complex mixture and because it illustrates many of the problems that attend characterization of toxic substances: There is a sparse database on the mixture and on many of its individual components, composition varies between lots, and there are few data on human exposure. Because of the wide range of environmental conditions of human exposure (e.g., extreme cold to extreme heat, variable humidity), the actual exposure to aerosolized or vaporized components of the fuel varies with the environmental circumstance. The subcommittee evaluated the toxicity of JP-8 only under standard conditions. A complete assessment by the Navy would require an evaluation of each component under the full range of environmental conditions in which human exposures occur. A number of toxicity studies, including reproductive and developmental toxicity studies, have been conducted on HFC-134a. HFC-134a was selected because more data are available for this compound than for JP8.

Additional examples of the application of the evaluative process to specific agents can be found in the literature (Moore et al. 1995b; 1997); *An Assessment of Boric Acid and Borax Using the IEHR Evaluative process For Assessing Human Developmental and Reproductive Toxicity of Agents.* Reproductive Toxicology, 11(1): 123-160; and *An Assessment of Lithium Using the IEHR Evaluative process For Assessing Human Developmental and Reproductive Toxicity of Agents.* Reproductive Toxicology, 9(2):175-210.

JP-8 JET FUEL

Jet fuel JP-8 (jet propellant-8) is a kerosene-based distillate selected by the U.S. Air Force to replace JP-4 and other predecessors, which were replaced because JP-8 has a higher flash point, is composed of higher chain hydrocarbons, and does not contain benzene. Profiles for JP-8 list the following classes of compounds exclusive of additives: alkanes (43% by weight); cycloalkanes (11%); alkylbenzenes (12%); naphthalenes (2%); and dicycloparaffins, tetralins, and olefins (% not specified) (USAF 1991). A more detailed list of hydrocarbon components is given in Table A-1.

Another jet fuel, JP-5, is physically and chemically similar to JP-8, and the differences between these fuels are considered minor (ATSDR 1998). Several studies described below were conducted using JP-5.

Exposure Data

Human exposure to JP-8 occurs during refueling and defueling operations and during mechanical activities that deal with storage, transfer, and combustion. Military personnel can be exposed to JP-8 by the inhalation (of aerosolized or vaporized fuel), dermal, and oral routes of exposure.

Occupational standards for JP-8 are primarily based on knowledge about the toxicity of kerosene and naphtha (a petroleum distillate fraction). National Institute for Occupational Safety and Health (NIOSH) guidelines include an 8-hour (hr) time-weighted-average recommended exposure limit (TWA-REL) for naphtha of 400 milligrams per cubic meter (mg/m^3) (100 parts per million (ppm)) (NIOSH

TABLE A-1 Composition of JP-8 Jet Fuel

Hydrocarbon Type	Weight % JP-8[a]
Isooctane	3.66
Methylcyclohexane	3.51
m-Xylene	3.95
Cyclooctane	4.54
Decane	16.08
Butylbenzene	4.72
1,2,4,5-Tetramethylbenzene	4.28
Tetralin	4.14
Dodecane	22.54
1-Methylnapthalene	3.49
Tetradecane	16.87
Hexadecane	12.22

[a] Composition of surrogate JP-8 (USAF 1991).

1999). Naval Occupational Safety and Health recommends a permissible exposure limit (PEL) of 350 mg/m^3 and a 15-minute (min) short term exposure limit (STEL) of 1,000 mg/m^3 (D.T. Harris et al. 1997). Puhala et al. (1997) reported measurement of jet fuel vapors at three domestic Air Force installations. Breathing-zone samples were collected from workers involved in aircraft maintenance, fuel handling, and flightline positions. Exposures at the base that used only JP-8 are listed in Table A-2. Each exposure fell below the American Conference of Governmental Industrial Hygienists (ACGIH) TWA threshold limit values (TLVs) for the chemicals analyzed. Two recent studies measured exposure of Air Force personnel to jet fuels, including JP-8. Pleil et al. (2000) used newly developed methods to collect exhaled breath from personnel at Air Force bases and then analyzed the samples for certain volatile marker compounds for JP-8 and for aromatic hydrocarbons such as benzene. The study authors found a demonstrable JP-8 exposure for all subjects, ranging from slight elevations to greater than 100-fold when compared with a control cohort. Carlton and Smith (2000) collected breathing zone samples from workers

TABLE A-2 Mean Exposure Concentrations of Jet Fuel Vapors, ppm[a]

Analyte	Overall Mean (n)		Standard Deviation	ACGIH TWA-TLV
Naphthas	0.359	(26)	.556	300
Benzene	0.003	(26)	.003	10[b]
Heptane	0.003	(26)	.006	400
m-Xylene	0.005	(26)	.008	100
o-Xylene	0.003	(26)	.004	100
p-Xylene	0.004	(26)	.005	100
Toluene	0.006	(26)	.012	50

[a]From Puhala et al. (1997) for base A where only JP-8 was used.
[b]The Occupational Safety and Health Administration PEL for benzene is 1 ppm.

during aircraft fuel tank entry and repair at 12 Air Force bases. They report that the highest 8-hr time-weighted average fuel exposure found was 1,304 mg/m^3, and the highest 15-min short-term exposure was 10, 295 mg/m^3.

General Toxicological and Biological Parameters

Lethality

Several case studies have reported death following accidental ingestion of kerosene by children (reviewed in ATSDR 1998). The primary cause of death is respiratory effects (lipoidal pneumonia). The lowest dose of kerosene associated with death was 1,900 milligrams per kilogram (mg/kg) body weight by a 2-year-old child. Doses ranging from 120 to 870 mg/kg and as high as 1,700 mg/kg did not lead to death in children ranging from 10 months to 5 years old.

No studies have reported death in humans associated with inhalation or dermal exposure to kerosene-based fuels.

The acute oral lethal dose for 50% of the test animals (LD$_{50}$) of JP-5 in rats (Bogo et al. 1983) and of kerosene in guinea pigs and rabbits (Deichmann et al. 1944) is greater than 10 grams (g) per kg.

Acute Studies

Acute exposure to kerosene-based fuels, such as JP-8, has been associated with respiratory, cardiovascular, ocular, neurological, immunological, renal, and dermal effects. Those studies are briefly described below and are summarized in Table A-3.

Human Studies

After inhalation of JP-5 for approximately 1-hr, two individuals experienced mild hypertension, eye irritation, and neurological effects (e.g., coordination and concentration difficulties, fatigue, headache, apparent intoxication, anorexia) and one individual experienced nausea (Porter 1990). All symptoms subsided by 4 days (d) after exposure. The concentration of JP-5 was not known. Six volunteers exposed to kerosene vapor at 140 mg/m³ for 15 min did not experience any respiratory effects (Carpenter et al. 1976).

Ingestion of kerosene by children and adults has been reported to cause pulmonary (e.g., pneumonia, bronchitis) and neurological (e.g., unconsciousness, semiconsciousness, drowsiness, restlessness, irritability) effects, tachycardia, cardiomegaly, vomiting, and increased leukocyte counts (reviewed in ATSDR 1998). Because in many cases ingestion of kerosene is accidental, the concentrations associated with specific effects are not reported. It has been estimated that respiratory distress will result from ingestion of 10-30 milliliters (mL) of kerosene (Zucker et al. 1986). Neurological effects (e.g., convulsions, coma) were observed in 2 of 78 children ingesting approximately 30 mL of kerosene; those effects were not observed in children ingesting from 3 to 20 mL.

There are no studies assessing acute dermal exposure in humans.

Experimental Animal Studies

Respiratory effects, such as bronchoconstriction, were observed in rabbits and guinea pigs exposed by inhalation to kerosene (Casaco et al. 1982; Garcia et al. 1988). The rabbits were exposed to 32,500 mg/m³ for 4-9 min, and the guinea pigs were exposed to 20,400 mg/m³ for 5 min. A study exposing mice to 20 microliters (μL) of kerosene (the only dose tested) by aspiration reported that the animals showed

TABLE A-3 Experimental Animal Studies on the Toxicity of Kerosene and Kerosene-Based Fuels

Study Type	Fuel Type	Species	Exposure Concentration, Duration, Route	Observed Effect	Reference
Acute	Kerosene	Rabbit	32,500 mg/m³, 4-9 min, inhalation (aerosol)	Reduction in tidal volume and dynamic lung compliance, bronchoconstriction, increase in pulmonary resistance	Casaco et al. 1982
Acute	Kerosene	Guinea pig	20,400 mg/m³, 5 min, inhalation (aerosol)	Bronchoconstriction	Garcia et al. 1988
Acute	Kerosene	Mouse	20 μL aspiration	Pulmonary consolidation and hemorrhage, pneumonitis, decrease in pulmonary clearance of *S. aureus*, increase in relative lung weight, neurological effects including lack of coordination, drowsiness, behavioral changes	Nouri et al. 1983
Acute	Kerosene	Guinea pig	3,200-8,000 mg/kg, gavage	Mononuclear and polymorphonuclear cell infiltration and unspecified pathological lesion in the lungs	J. Brown et al. 1974
Acute	Kerosene	Dog	0.5 mL/kg, aspiration	Increases in arterial oxygen utilization, intrapulmonary physiological shunt fraction, respiratory rate. Decreases in arterial oxygen tension, heart rate, mean arterial blood pressure	Goodwin et al. 1988

Duration	Fuel	Species	Dose/Route	Effects	Reference
Acute	Kerosene	Rat	8,000 - 12,000 mg/kg, gavage	Histopathological changes in kidneys (no change in kidney weight). Neurological effects such as unsteady gait and drowsiness observed at 12,000 mg/m^3, but not 8,000 mg/m^3.	Muralidhara et al. 1982
Acute	JP-5	Rat	18,912 mg/kg, gavage	Hematological, hepatic, renal effects	Parker et al. 1981
Acute	JP-5	Rat	24-60 mL/kg, gavage	Hepatic effects	Bogo et al. 1983; Mehm and Feser 1984
Acute	JP-5	Rat	19,200 mg/kg, gavage	Renal effects such as hyaline droplets in cytoplasm of epithelial cells in proximal tubules; neurological effects such as reduction in food, water intake; dermal effects such as alopecia, congestion of the subcutis	Bogo et al. 1983
Acute	JP-5 or JP-8	Rabbit	0.5 mL, dermal (undiluted fuel)	No effects observed	Schultz et al. 1981
Acute	JP-8	Rabbit	Dermal, abraded and intact skin	Slight skin irritation	Kinkead et al. 1984
Acute	JP-5	Mouse	Concentration not reported, dermal	Dermatitis observed	NTP/NIH 1986
Acute	JP-5	Guinea pig	1% solution, dermal	Mild dermal sensitization	Cowan and Jenkins 1981a,b

TABLE A-3 *(Continued)*

Study Type	Fuel Type	Species	Exposure Concentration, Duration, Route	Observed Effect	Reference
Subchronic	Kerosene	Rat and dog	100 mg/m³, 6 hr/d, 5 d/wk for 13 wk, inhalation	No respiratory, cardiovascular, gastrointestinal, hematological, musculoskeletal, hepatic, renal, body weight, neurological effects observed	Carpenter et al. 1976
Repeated	Kerosene	Rat	1,100 mg/m³, 6 hr/d, 5 d/wk for 30 d, inhalation	No hepatic effects observed	Bogo et al. 1983
Repeated	JP-8	Rat	497 mg/m³, 1 hr/d for 7 or 28 d, inhalation (nose only)	Increased alveolar permeability	USAF 1994; Chen et al. 1992
Repeated	JP-8	Rat	500-1,100 mg/m³, 1 hr/d for 7, 28, 56 d, inhalation	Lung epithelial permeability observed in rats exposed for 56 d	USAF 1994; Hays et al. 1994
Repeated	JP-8	Rat	950 mg/m³, 1 hr/d for 28 d, inhalation	Disruption of epithelial and endothelial structures, convoluted airways, and alveoli filled with red blood cells and fluid	USAF 1994; Pfaff et al. 1993
Repeated	Kerosene	Guinea pig	20,400-34,000 mg/m³, 15 min/d for 21 d, inhalation	Cardiovascular effects (aortic plaques) observed	Noa and Illnait 1987a,b

Subchronic	JP-5	Rat and Dog	150-750 mg/m^3, continuous exposure for 90 d, inhalation	Female rats exposed at 150 or 750 mg/m^3 and male rats exposed at 750 mg/m^3 showed increased creatinine and blood urea nitrogen; dogs exposed at 750 mg/m^3 had a slight (statistically significant) decrease in hemoglobin and red blood cell count, significant decreases in serum albumin, sporadic changes in blood urea nitrogen; dogs exposed at 150 or 750 mg/m^3, showed hepatic effects such as lesions, mild cloudy swelling of hepatocytes	USAF 1978b
Repeated	Kerosene	Rat	Average concentrations of 58 mg/m^3 and 231 mg/m^3, "intermediate duration exposure" inhalation	At 58 mg/m^3, animals had decreased blood glucose; at 231 mg/m^3 had increased blood lactate, pyruvate	Starek and Vojtisek 1986
Subchronic	JP-5	Mouse	150 mg/m^3, continuous exposure for 90 d, inhalation	Vacuolization, hepatocellular fatty changes observed in the livers	Gaworski et al. 1984
Subchronic	JP-8	Rat and mouse	500 mg/m^3 and 1,000 mg/m^3, continuous exposure for 90 d, inhalation	Kidney lesion, α-2-microglobulin protein droplet nephropathy, observed in male rats. No exposure-related increase in either sex of either species. Condition not considered relevant to humans	Mattie et al. 1991

TABLE A-3 (*Continued*)

Study Type	Fuel Type	Species	Exposure Concentration, Duration, Route	Observed Effect	Reference
Subchronic	JP-8	Rat	750, 1,500, 3,000 mg/m^3/d for 90 d, gavage	Nephropathy, dose-related decrease in body weight, and dose-related increase in gastritis and perianal dermatitis	Mattie et al. 1995
Repeated	JP-8	Mouse	100, 250, 500, 1,000, 2,500 mg/m^3, 1 hr/d for 7 d, inhalation (aerosolized)	Dose-related decrease in spleen and thymus organ weights and decrease in total viable cells from those organs. At 100 and 250 mg/m^3, loss of total cell numbers in the lymph nodes and peripheral blood, but an increase in bone marrow total cell numbers; at 500 and 1,000 mg/m^3, increase in total cell numbers in lymph nodes and peripheral blood cell numbers, but a decrease in bone marrow cell numbers. Dose-related decrease in immune function observed in animals exposed to JP-8 and then treated with Concavalin-A	D.T. Harris et al. 1997
Repeated	Kerosene	Mouse	0.1 mL/d for 1 wk, dermal	Dermal (rough skin, edema, inflammation, dermatosis), hematological, immunological (decreases in relative weight of lymph nodes and thymus and decreases in thymus counts, bone marrow nucleated cell counts, thymic cortical lymphocytes, and cellularity of the thymic lobules), and behavioral (tactile stimuli and hyperactivity) effects.	Upreti et al. 1989

Study	Compound	Species	Dose/Exposure	Effect	Reference
Repeated	JP-5	Mouse	50% or 100% 3 times/wk for 60 wk, dermal	Renal lesions	Easley et al. 1982
Repeated Subchronic	JP-5	Mouse	2,500–8,000 mg/kg, for 5 d/wk for 13 or 103 wk, dermal	Dermal (rough skin, edema, inflammation, dermatosis) and hematological effects. Hepatic effects observed in the 13-wk study. Animals treated at 8,000 mg/kg had slightly decreased body weight. Granulocytic hyperplasia in bone marrow and hyperplasia in lymph nodes.	NTP/NIH 1986
Carcinogenicity	JP-5	Rat	750 mg/m^3, continuous exposure for 90 d, inhalation; animals followed for their lifetime	No renal tumors	Bruner 1984
Carcinogenicity	JP-5	Mouse	22.9 mg and 42.9 mg for 40 wk, dermal	Skin tumors (type not reported) in animals exposed at 22.9 mg but not at 42.9 mg	Schultz et al. 1981
Genotoxicity	JP-8	Salmonella typhimurium	0.001 µL/plate to 5.0 µL/plate	Not mutagenic in 5 strains of S. typhimurium with or without metabolic activation	USAF 1978a
Genotoxicity	JP-8	L5178Y Mouse lymphoma cells	0.01 µL/mL to 0.16 µL/mL	Gene mutations not induced in lymphoma cells at thymidine kinase locus	USAF 1978a
Genotoxicity	JP-8	WI-38 cells	5.0 µL/mL	JP-8 induced significant increases in ^{3}H-thymidine incorporation in WI-38 cells	USAF 1978a

TABLE A-3 (*Continued*)

Study Type	Fuel Type	Species	Exposure Concentration, Duration, Route	Observed Effect	Reference
Genotoxicity	JP-8	Rat and mouse	Rat: 0.1 mL/kg, 0.3 mL/kg, 1.0 mL/kg; mouse: 0.13 mL/kg, 0.4 mL/kg, 1.3 mL/kg; in the feed for 5 d	JP-8 negative in dominant lethal assays	USAF 1978a
Developmental toxicity	JP-8	Rat	500, 1,000, 1,500, 2,000 mg/kg/d, given on d 6 to 15 of pregnancy, oral	Maternal weight gain decreased significantly at 1,000 mg/kg/d. Several deaths among animals exposed to JP-8 were attributed to the presence of JP-8 in the lungs. At 1,500 and 2,000 mg/kg/d, fetal weight was reduced by 12% and 25% and maternal weight gain was reduced by 70% and 85%, respectively. Number and type of fetal malformations and variations observed did not differ significantly between dose groups. Progressive increase in the overall incidence of fetal alterations reported between 500 and 1,500 mg/kg/d, but not 2,000 mg/kg/d. The 2,000 mg/kg/d group had fewer fetuses available for examination than other groups because approximately 1/3 of the dams died. One animal had totally resorbed litter.	Cooper and Mattie 1996

respiratory effects (e.g., pulmonary consolidation and hemorrhage, pneumonitis, a decrease in pulmonary clearance of *Staphylococcus aureus*, an increase in relative lung weight) and neurologic effects (e.g., lack of coordination, drowsiness, behavioral fraction, and respiratory rate and decreases in arterial oxygen tension, heart rate, and mean arterial blood pressure (Goodwin et al. 1988). Rats exposed to kerosene administered by gavage at single doses of up to 12,000 mg/kg were not found to exhibit cardiovascular, gastrointestinal, hematological, hepatic, or endocrine effects (Muralidhara et al. 1982). However, they did exhibit histopathological kidney changes (although no change was observed in kidney weight) and some neurological effects, including unsteady gait and drowsiness. No neurological effects were observed in rats exposed at 8,000 mg/kg. Rats exposed by gavage to a single dose of JP-5 at 18,912 mg/kg did have hematological, hepatic, and renal effects (Parker et al. 1981). Hepatic effects were observed in rats exposed by gavage to single doses of JP-5 at 24-60 mL/kg (Bogo et al. 1983; Mehm and Feser 1984). Renal (e.g., hyaline droplets in the cytoplasm of epithelial cells in the proximal tubules), neurological (e.g., reduction in food and water intake), and dermal (e.g., alopecia and congestion of the subcutis) effects were found in rats exposed to a single dose by gavage at 19,200 mg/kg of JP-5 (Bogo et al. 1983).

Rabbits exposed dermally to undiluted JP-5 or JP-8 at 0.5 mL did not show any signs of dermal effects (Schultz et al. 1981); however, in another study using rabbits dermally exposed to JP-8, slight skin irritation was observed (Kinkead et al. 1984). Also, dermal effects were observed in mice exposed to JP-5 (concentration not reported) (NTP/NIH 1986). Acute dermal exposure of mice to JP-5 at 10,000 mg/kg led to decreased body weight, but exposure at 5,000 mg/kg did not have this effect. Acute dermal exposure of guinea pigs to a 1% solution of JP-5 led to mild dermal sensitization (Cowan and Jenkins 1981a).

Repeated-Dose Studies

Repeated exposure to kerosene-based petroleum distillates, such as JP-8, has been associated with hepatic, renal, cardiovascular, neurological, and pulmonary toxicity in humans and experimental animals. These studies are described briefly and summarized in Table A-3.

Human Studies

Struwe et al. (1983) reported that airline industry workers occupationally exposed by inhalation, oral, and dermal routes to jet fuels (type not specified) were examined for neuropsychiatric effects. Thirty employees exposed at an estimated time-weighted average of 250 mg/m^3 during work for 4 to 32 years were examined. The study authors concluded that personality changes and emotional dysfunctions are effects of long-term exposures to jet fuels. The usefulness of this study for determining the general toxicity of exposure to JP-8 in the context of Naval operations is unclear for the following reasons: (1) the exposed workers studied were not a random sample but "were selected in collaboration with the management of the factory, the trade unions and the health department of the factory" and the criteria for selection are not reported; (2) the composition of the jet fuel or fuels involved is not reported, and their similarity to JP-8 is uncertain; (3) the average exposure for these workers was estimated to be in the range of the permissible exposure limit, however, 21 of the 30 workers are reported to have had recurrent acute exposures that produced symptoms such as dizziness, headache, nausea, palpitations, and feelings of suffocation; and (4) although the medical history obtained did not indicate that the exposed workers were more likely than controls to have selected confounding factors, possible confounders were not otherwise considered in the data analysis.

In an epidemiological study reported by Knave et al. (1978), factory workers chronically exposed by inhalation, oral, and dermal routes to jet fuel (fuel type not reported) were found to have significant increases in conditions such as fatigue, depression, dizziness, and sleep disturbances. Also, the workers reported a significant increase in "a feeling of heaviness" in their chests. An estimated time-weighted average of 128-423 mg/m^3 jet fuel was found in the work area. A limitation of this study is that other exposures were not considered. Factory workers chronically exposed by the dermal route to kerosene for up to 5 hr/d exhibited dermatosis and erythema (Jee et al. 1985). The concentration of kerosene was not reported.

Experimental Animal Studies

Carpenter et al. (1976) exposed rats and dogs to deodorized kerosene at 100 mg/m³ for 6hr/d, 5d/wk, for 13 wk by inhalation. The animals did not exhibit any respiratory, cardiovascular, gastrointestinal, hematological, musculoskeletal, hepatic, renal, body weight, or neurological effects as a result of exposure. Hepatic effects also were not observed in a study in which rats were exposed by inhalation to 1,100 mg/m³ of JP-5 for 6 hr/d, 5d/wk, for 30 d (Bogo et al. 1983).

However, other studies have reported respiratory, cardiovascular, hematological, hepatic, renal, or body weight effects associated with exposure by inhalation to kerosene-based fuels. Rats exposed nose-only at 497 mg/m³ of JP-8 for 1hr/d for 7d or for 28 d showed increased alveolar epithelial permeability (Chen et al. 1992; USAF 1994). Lung epithelial permeability in rats was affected by exposure by inhalation of JP-8 at 500 and 800 to 1,100 mg/m³ for 56 d, but not for 7 or 28 d (USAF 1994; Hays et al. 1994). Rats exposed to an average concentration of 950 mg/m³ of JP-8 for 28 d exhibited pathological changes, including disruption of epithelial and endothelial structures, convoluted airways, and alveoli filled with red blood cells and fluid (Pfaff et al. 1993; USAF 1994).

Cardiovascular effects (aortic plaques) were observed in guinea pigs exposed by inhalation to kerosene at 20,400-34,000 mg/m³ for 15 min/d for 21 d (Noa and Illnait 1987a,b). Hematological effects were observed in dogs and rats exposed by inhalation to JP-5 (USAF 1978b). The rats were exposed at 150 or 750 mg/m³ (females) and 750 mg/m³ (males) for 90 d. The dogs were exposed at 750 mg/m³ for 90 d. Changes in blood glucose, blood lactate, and pyruvate concentrations were observed in rats exposed by inhalation at an average of 58 mg/m³, 231 mg/m³, and 231 mg/m³, respectively (Starek and Vojtisek 1986). Dogs exposed to JP-5 at 150 or 750 mg/m³ for 90 d showed hepatic effects, including lesions, and mild cloudy swelling of hepatocytes (USAF 1978b). The nature of those lesions was not reported. Hepatic changes also were observed in mice exposed to JP-5 at 150 mg/m³ for 90 d (Gaworski et al. 1984).

In a study by Mattie et al. (1991), Fischer 344 rats and C57Bl/6 mice were exposed to vapors of JP-8 at 0, 500, or 1,000 mg/m³ continuously for 90 d and were held for further observation as long as 21 months. The only toxicity observed was a kidney lesion, α-2u-globulin protein droplet nephropathy, specific to male rats. There was no exposure-related increase in the incidence of tumors in either sex of either species. However, it should be noted that a 3-month exposure period is not generally considered adequate for a rigorous evaluation of carcinogenic potential. In a subsequent study, male Sprague-Dawley rats were dosed with JP-8 at 0, 750, 1,500, 3,000 mg/kg/d by gavage for 90 d to further characterize kidney lesion and assess further toxic effects (Mattie et al. 1995). In addition to the α-2u-globulin protein droplet nephropathy observed in male rats, there was a dose-related decrease in body weight, a dose-related increase in gastritis and perianal dermatitis, and an increase in liver enzymes that was not related to the dose of JP-8. Several other studies have reported this type of nephropathy in rats treated via inhalation with jet fuels at 150 or 750 mg/m^3 (Cowan and Jenkins 1981a,b; Bruner 1984; Gaworski et al. 1984; USAF 1985). Because the condition is specific to male rats, it is not relevant to humans.

D.T. Harris et al. (1997) exposed C57Bl/6 mice to aerosolized JP-8 for 1 hr/d for 7 d at 0, 100, 250, 500, 1,000, 2,500 mg/m^3 to determine possible immunotoxicity. Dose-related immunological effects seen at the lowest concentration (100 mg/m^3) included a decrease in spleen and thymus organ weight and a decrease in total viable cells recovered from those organs. At low exposure concentrations (100 and 250 mg/m^3), there was a loss of total cell numbers in the lymph nodes and peripheral blood; at higher exposure concentrations (500 and 1,000 mg/m^3), there was an increase in total cell numbers. Bone marrow analysis showed that exposure to low concentrations resulted in an increase of total cell numbers and that exposure to higher concentrations resulted in a decrease in total cell numbers. To determine whether exposure to JP-8 can cause loss of immune function, splenic immune cells were examined for the ability to undergo functional responses after stimulation by a growth factor and a mitogen. Dose-related decreases in immune function were observed in mice exposed to JP-8 and stimulated with the T-cell mitogen, Concavalin-A. The authors concluded that at concentrations as low as 100 mg/m^3, JP-8 can act as an immunosuppressive agent.

No studies have evaluated the toxicity of kerosene-based fuels as a result of multiple oral exposures.

Two studies tested the toxicity in mice exposed by dermal administration of kerosene at 0.1 mL/d for 1 wk (Upreti et al. 1989) and JP-5 at 2,500 to 8,000 mg/kg for 5 d/wk for 13 wk and 103 wk (NTP/NIH 1986). Hematological effects were observed at all concentrations and durations tested, and hepatic effects were observed at all concentrations in the 13-wk study. Dermal effects (rough skin, edema, inflammation, dermatosis) in mice were reported in both studies. No respiratory, cardiovascular, gastrointestinal, musculoskeletal, renal, or endocrine effects were reported in either study. Renal lesions were reported in mice exposed dermally to 100% or 50% JP-5 for 3 times/wk for 60 wk (Easley et al. 1982). Mice treated dermally with JP-5 at 8,000 mg/kg showed small (3-7%) changes in body weight (NTP/NIH 1986). In the study by Upreti et al. (1989), male mice (females were not tested) treated with JP-5 showed decreases in relative lymph node and thymus weight and decreases in thymocyte count, bone marrow nucleated cell count, thymic cortical lymphocytes, and the cellularity of the thymic lobules. The NTP/NIH (1986) study found induced granulocytic hyperplasia in the bone marrow and hyperplasia in the lymph nodes of mice treated with JP-5. Male mice treated with JP-5 dermally to 0.1 mL of kerosene per day for 1 wk had increased response to tactile stimuli and hyperactivity (Upreti et al. 1989).

Cancer

No epidemiology studies have been conducted to determine the carcinogenicity of JP-8 or other kerosene-based fuels.

No oral carcinogenicity studies have been conducted in experimental animals exposed to kerosene, JP-5, or JP-8. No renal tumors were observed in rats after continuous exposure to JP-5 at 750 mg/m^3 for 90 d and followed for their lifetime (Bruner 1984). Skin tumors (type not reported) were observed in mice exposed dermally at 22.9 mg of JP-5 for 40 wk; however, tumors were not observed at a dose of 42.4 mg (Schultz et al. 1981).

Genetic Toxicity

The genotoxic potential of JP-8 has been evaluated in a battery of tests (USAF 1978a). JP-8 was not mutagenic in five strains of *Salmonella typhimurium* with or without metabolic activation. Gene mutations were not induced in L5178Y mouse lymphoma cells at the thymidine kinase locus. In a test for unscheduled DNA synthesis, JP-8 induced significant increases in ^{3}H-thymidine incorporation in WI-38 cells at 5.0 μL/mL. JP-8 was negative in dominant lethal assays in mice and rats. The mice were administered the test compound in the feed for 5 d at concentrations of 0.13 mL/kg, 0.4 mL/kg, and 1.3 mL/kg. The rats were exposed via the same route and duration at concentrations of 0.1 mL/kg, 0.3 mL/kg, and 1.0 mL/kg.

Disposition and Pharmacokinetics

Because JP-8 is a complex mixture of numerous volatile hydrocarbons and other substances, it is difficult to describe the pharmacokinetics both of the mixture and of its components as they relate to toxicity. The pharmacokinetics of some JP-8 components are known, but the usefulness of such data is limited because some components of the mixture likely affect the kinetics of uptake, distribution, metabolism, and elimination of others in the mixture. The kinetics of the mixture also would vary by route (e.g., oral versus dermal) and condition of exposure (e.g., aerosol versus vapor).

Riviere et al. (1999) used the isolated perfused porcine skin flap model to study absorption and disposition of JP-8. The percutaneous absorption and cutaneous disposition of topically applied neat Jet-A and JP-8 jet fuels were assessed by monitoring the absorptive flux of the marker components ^{14}C naphthalene and ^{3}H dodecane simultaneously. Absorption of ^{14}C hexadecane was estimated from JP-8. Data were not reported in absolute amounts or concentrations. Instead, the objectives were to determine the relative absorption of the individual marker components from jet fuel, and the effect of a specific jet fuel's composition on the absorption of a specific marker. Having evaluated the absorption of only three of the 228 major nonadditive hydrocarbon constituents of the fuels, the authors stated that this is insufficient information to conduct risk assessments on jet fuels. However, the authors' conclusions are informative. Naphthalene penetrated the skin more rapidly than dodecane or hexadecane, but the latter compounds had a larger fraction of the dose deposited in the skin. There were also differences in naphthalene and dodecane absorption and skin deposition between the fuels. These findings reinforce the difficulty of predicting risk for complex mixtures such as jet fuels.

Reproductive and Developmental Toxicity Data

Human Studies

Two studies have been published about possible genotoxicity and male reproductive toxicity in aircraft maintenance personnel exposed to solvents, paints, and fuels (mainly JP-4; Lemasters et al. 1999a,b). A total of 50 men working on aircraft maintenance at an Air Force base were included in the studies. The subjects were divided into subgroups based on work assignment, and therefore related chemical exposure: 6 sheet metal workers, 6 painters, 15 men involved in jet fueling, and 23 flightline workers. Eight unexposed men served as a control group. All measures of chemical exposure were below 6 ppm, well below the Occupational Safety and Health Administration standards for those chemicals. Evaluation of blood lymphocytes for genotoxic changes after 15 and 30 wk exposure as measured by sister chromatid exchanges and micronuclei revealed no significant changes in either parameter among the jet fueling and flight line groups of men (Lemasters et al. 1999a). The reproductive study included measures of sperm production, structure, and function (sperm concentration, sperm motion, viability, morphology, morphometrics, and stability of sperm chromatin) after 15 and 30 wk exposure. There was an increase in sperm concentration in the jet fuel and flightline groups and a decrease in sperm linearity in the jet fuel group, but the authors concluded that exposure to jet fuel did not cause an apparent effect on semen quality for aircraft maintenance personnel (Lemasters et al. 1999b).

No human studies have been conducted to assess female reproductive or developmental toxicity caused by exposure to JP-8 or any other kerosene-based fuel.

Experimental Animal Studies

Developmental Toxicity

Cooper and Mattie (1996) reported the results of a study of the developmental toxicity of JP-8 in Sprague-Dawley rats dosed orally at 0, 500, 1,000, 1,500, 2,000 mg/kg/d on days 6-15 of pregnancy. Dams exposed to doses of 1,000 mg/kg/d or above gained significantly less body weight during pregnancy than did control rats. There were several maternal deaths among exposed animals that were attributed to the presence of JP-8 in the lungs. Fetal body weight at the two highest doses was significantly decreased from control weight, but those doses were associated with even greater reduction of maternal weight gain during pregnancy. Fetal weight was reduced by 12% and 25%, and maternal gestational weight gain was reduced by 70% and 85% at concentrations of 1,500 and 2,000 mg/kg/d, respectively. It is unclear if the fetal weight reduction was causally associated with reduced maternal gestational weight gain.

The number and type of fetal malformations and variations observed did not differ significantly between dose groups. A progressive increase in the overall incidence of fetal alterations (variations and malformations) with increasing dose was reported between the 500 mg/kg/d and 1,500 mg/kg/d dose groups, but not for the 2,000 mg/kg/d dose group. It should be noted that the number of fetuses and litters exposed to the highest dose (2,000 mg/kg/d) and available for examination for abnormal development was much lower than in other dose groups because approximately one-third of the dams died; one animal had a totally resorbed litter. Observed variations included dilated renal pelvis, ureter, and lateral ventricle; unossified sternebra; rudimentary 14th rib; less than four metatarsals; and external and subdural hematomas. Observed malformations included malformed sternum, missing centrum, hydronephrosis, ectopic heart, short tail, no tail, and encephalomyocoele.

No other studies have assessed the developmental toxicity of JP-8 (or other kerosene-based fuels) in experimental animals.

Reproductive Toxicity

No studies of reproductive toxicity of JP-8 (or other kerosene-based fuels) in experimental animals were found. Ancillary data from other toxicity studies do not suggest an adverse reproductive effect (no effect on fertility in dominant lethal studies in mice and rats (USAF 1978a) and no effect on testis weight or histopathology in a 90-d gavage study in rats (Mattie et al. 1995)). An increase in atrophy of seminiferous tubules in testes of male mice exposed to JP-4 by inhalation for 12 months was considered by the authors to result from the debilitating effects of chronic skin disease in exposed mice (Bruner et al. 1993).

Integration of Toxicity and Exposure Information

Interpretation of Toxicity Data

Data to assess the potential of JP-8 to adversely affect reproduction and development are sparse. One study (Puhala et al. 1997) reported measurements of human exposures and the values for the components of jet fuels analyzed that were far below the TWA threshold limit values (see Tabel A-2). Data on the absorption of volatile hydrocarbon components of JP-8 suggest that systemic exposure is likely, by any route of exposure. The single published developmental toxicity study (Cooper and Mattie 1996) did not report an adverse effect on embryonic or fetal development in rats with oral treatment at up to 2,000 mg/kg/d on days 6-15 of pregnancy, except for a decrease in body weight of offspring.

No studies of humans or experimental animals have been done to assess reproductive performance after exposure to JP-8. There are human data that demonstrate that exposure to jet fuel (mostly JP-4) at below 6 ppm did not affect semen quality for aircraft maintenance personnel (Lemasters et al. 1999b). Ancillary studies in rats and mice (USAF 1978b; Mattie et al. 1995) did not suggest an adverse effect on reproductive organs or reproductive performance. The testicular atrophy reported in mice exposed to JP-4 (Bruner et al. 1993) might have been secondary to the debilitating effect of chronic skin disease.

Quantitative Evaluation

One study identified a NOAEL (no-observed-adverse-effect level) for a reproductive or developmental endpoint (Cooper and Mattie 1996). Rats were exposed orally at 0, 500, 1,000, 1,500, and 2,000 mg/kg/d on days 6-15 of pregnancy. There was a significant decrease in fetal body weight in rats exposed to JP-8 at high doses (1,500 and 2,000 mg/kg/d). No effect on fetal body weight was observed after exposure at 1,000 mg/kg/day, and the authors identified that dose as the fetal NOAEL.

Maternal toxicity in the form of death and decreased body weight occurred at the same doses at which developmental toxicity (decreased fetal body weight) was observed. No effect on maternal body weight was observed after exposure at 500 mg/kg/d, and that dose was identified as the maternal NOAEL. Whether the developmental toxicity is directly related to the maternal toxicity or is independent of the effects on the mothers is not known.

Data on comparative pharmacokinetics are sparse; there are no data to support a conclusion that adverse reproductive or developmental toxic effects in rats or mice are not predictive of some adverse effect in humans. Thus, it is accepted by default that animal data are relevant to humans.

Estimates of human exposure to JP-8 do not provide documentation of exposures to individual components of JP-8. Studies have not been done to determine which components of JP-8 might account for its toxicity. Thus, the value of calculation of a margin of exposure for JP-8 is questionable because knowledge of the composition of JP-8 might not accurately predict the relative exposure to components of JP-8 at the tissue level.

The subcommittee chose the NOAEL of 1,000 mg/kg/d to calculate an unlikely effect level (UEL) for developmental toxicity. The aggregate uncertainty factor for human sensitivity is 1,000 (10 for interindividual variation, 10 for extrapolation from rats to humans, 10 for an incomplete data set). The UEL is calculated by dividing the NOAEL by the aggregate uncertainty factor for human sensitivity:

$$\text{UEL for developmental toxicity} = \frac{\text{NOAEL (1,000 mg/kg/d)}}{1,000} = 1 \text{ mg/kg/d}.$$

The UEL is only for effects that are observed at birth and only for a short term exposure. No long-term follow-up studies (e.g., on neurotoxicity) have been conducted. UELs for other reproductive endpoints cannot be calculated. A UEL for chronic exposure to JP-8 was not calculated because there are no chronic toxicity studies on JP-8 reported in the literature. Conversion from mg/kg/day by the oral route to the equivalent concentration in inhaled air to achieve the same daily dose determines that the 1 m/kg/d UEL is equivalent to 1.5 ppm for rats (assume 8 hr/d exposure, 100% absorption, 185 g body weight, respiratory minute volume of 0.76 mL/min/g body weight), and 0.8 ppm for humans (assume 8 hr/d exposure, 100% absorption, 69 kg body weight, respiratory minute volume of 0.42 mL/min/kg body weight).

Critical Data Needs

Data on the toxicity and disposition of JP-8 in animals are sparse, and no data are available for humans. No reproductive toxicity studies have been done in experimental animals. One adequate study demonstrated developmental toxicity in rats treated orally at 1,500-2,000 mg/kg/d (Cooper and Mattie 1996). A study in a second species should be supplemented with a multiple-generation reproductive toxicity study in rats or mice, including an evaluation of postnatal endpoints, such as developmental neurotoxicity, immunotoxicity, and hematological, hepatic, and renal effects, that could result from prenatal exposures.

Uncertainty about the toxicity of JP-8 could be reduced as follows:

- The toxicity of individual components of the fuel should be assessed from the literature to determine whether exposure to any of the components is known to produce reproductive or developmental toxicity in animals or humans.
- The pharmacokinetics of known components of the fuel should be assessed to better define exposure, placental transfer, bioaccumulation, and other factors relevant to toxicity.
- Research should determine systemic exposure from the various relevant modes of exposure. Toxicological studies have been done for gavage and inhalation of aerosolized fuel and vapors. Data are not available to determine the comparability of exposure by these routes and modes of exposure. Humans are exposed dermally as well as by inhalation. The contribution to the internal dose from exposure of the skin to JP-8 or its vapors is not known. Measurements of jet-fuel components in blood of exposed workers would permit comparisons to similar data

from laboratory animals and would facilitate extrapolation of toxicity data from animals to humans.

Additional data are needed to define exposures of humans and experimental animals better. Because JP-8 is a complex mixture of substances that differ in volatility, solubility, metabolic rates and pathways, and rate and route of elimination from the body, dosimetry of critical components of the mixture at critical sites in the body is crucial. Knowledge of the composition of JP-8 might not be a good surrogate for prediction of risk of some highly toxic minor component of the fuel.

Also, exposure to individual components of JP-8 under desert conditions of high temperature and low humidity would be different from exposures at very low temperatures because of different rates of aerosolization and vaporization. Exposure data should be collected from a variety of environmental conditions.

Summary

Jet propulsion fuel JP-8 is the fuel used by the U.S. Air Force and other services to fuel jets and other military vehicles. JP-8 is a mixture of hundreds of chemicals, mostly alkanes in the C8 to C17 range, and aromatics, including substituted benzenes and naphthalenes. The exact composition of JP-8 varies from batch to batch.

Human Exposure

Human exposure to JP-8 occurs during refueling and defueling operations and mechanical activities that deal with storage, transfer, and combustion. The most likely exposure of military personnel is via inhalation of aerosolized or vaporized fuel; however, topical and oral exposures also are possible. Occupational exposure standards are based on knowledge of the toxicity of components of JP-8. Those few exposure values that are published suggest that human exposures were below ACGIH TWA threshold limit values for those chemicals.

Toxicology

Developmental Toxicity

There are no human data on the effects of JP-8 on development. The animal data are sufficient to conclude that prenatal oral exposure at doses of 1,500 mg/kg/d and greater administered on gestation days 6-15 in rats causes developmental toxicity. These toxicity findings in rodents are assumed to be relevant for prediction of risk to humans.

Reproductive Toxicity

There are no human data on the effects of JP-8 on male or female reproduction. There are human data that show that exposure to jet fuel (mostly JP-4) at below 6 ppm did not affect semen quality for aircraft maintenance personnel. Likewise, there are no laboratory animal studies on the effects of JP-8 on male or female reproduction.

Quantitative Evaluation

Developmental Toxicity

There are no human data from which to develop a quantitative evaluation. One laboratory study in rats identified a NOAEL of 1,000 mg/kg for developmental effects. Using an aggregate uncertainty factor of 1,000 (10 for interindividual variation, 10 for extrapolation from rats to humans, 10 for an incomplete data set), the UEL for developmental toxicity for a short term exposure is 1 mg/kg/d.

Reproductive Toxicity

There are no human or animal data from which to develop a quantitative evaluation or calculate UELs for male or female reproductive toxicity endpoints.

Certainty of Judgment and Data Needs

Data on the toxicity and disposition of JP-8 in animals are sparse and no data are available for humans. One adequate study demonstrated developmental toxicity in rats. This study has not been replicated and there are no corroborative data from other studies with rats or other species. A multiple-generation reproduction study that examines a variety of postnatal endpoints that result from prenatal exposures, such as developmental neurotoxicity; immunotoxicity; and hematological, hepatic, and renal effects, should be conducted in rats or mice.

Additional data are needed to better define the exposure of humans and, in the context of animal toxicity studies, of laboratory animals. Because JP-8 is a complex mixture of chemicals that differ in volatility, solubility, metabolic rate and pathway, and rate and route of elimination from the body, dosimetry of critical components of the mixture at critical sites in the body is important to enhance the quality of risk assessment. The fact that human exposures can involve liquid fuel, aerosolized fuel, and vapor, by inhalation, dermal, and oral routes of exposure makes it difficult to accurately predict the internal dose of JP-8 and its components.

1,1,1,2-TETRAFLUOROETHANE[1]

Hydrofluorocarbons (HFCs), including 1,1,1,2-tetrafluoroethane (HFC-134a), have been developed as alternatives to chlorofluorocarbons (CFCs), which are known to contribute to the breakdown of ozone to oxygen in the stratosphere. HFCs do not contribute to the destruction of stratospheric ozone, but some HFCs have global warming potential. They primarily serve as replacements for CFCs in refrig-

---

[1]Subcommittee member Paul Foster was previously employed at a company that conducted reproductive and developmental toxicity studies on HFC-134a. Because Dr. Foster was involved in the review of those studies, he did not participate in the subcommittee's discussions and deliberations on HFC-134a.

eration equipment and mobile air conditioning; they also have pharmaceutical applications (e.g., as a propellant for metered dose inhalers used to treat asthma). The physical and chemical properties of HFC-134a are listed in Box A-1.

EPA has developed a chronic reference concentration (RfC) for chronic exposure of 80 mg/m^3 (Integrated Risk Information System (IRIS) 1998), based primarily on a 2-year inhalation study in rats (Collins et al. 1995). Briefly, male rats exposed at concentrations of 10,000 ppm and 50,000 ppm had a significant increase in the incidence of Leydig cell hyperplasia compared with controls. The study is described below.

The American Industrial Hygiene Association's (AIHA) Workplace Environmental Exposure Level Committee gave HFC-134a an occupational exposure limit (8-hr time-weighted average) of 4,250 mg/m^3 (AIHA 1991, as cited in European Centre for Ecotoxicology and Toxicology of Chemicals (ECETOC) 1995).

Exposure Data

Human exposure to HFC-134a occurs via inhalation from accidental leaks of air conditioning units and refrigerators, from spills or industrial use, and from use of metered-dose inhalers such as those that deliver medication for the treatment of asthma (Hazardous Substance Data Base (HSDB) 1998; Alexander and Libretto 1995).

The likely maximum exposure from a metered-dose inhaler is 33 ppm hr/m^3 lung surface area/d (Alexander et al. 1996).

General Toxicological and Biological Parameters

Acute Studies

No adverse health effects in humans from acute exposure to HFC-134a have been reported.

In experimental animals, HFC-134a has been shown to have low toxicity via inhalation. An approximate lethal concentration in rats ranges from 567,000 to 750,000 ppm after 4-hr and 30-min exposures,

Box A-1 Physical and Chemical Properties, HFC-134a

Common name:	FC-134a
Chemical name:	1,1,1,2-tetrafluoroethane
Synonyms:	HFC-134a; Norflurane; HFA-134a; 1,2,2,2-tetrafluoroethane; F-134a; R134a; Refrigerant R134a
CAS number:	811-97-2
Molecular formula:	C2-H2-F4
Description:	Colorless gas
Molecular weight:	102.03
Boiling point:	-26.5 °C at 736 mm Hg
Freezing point:	-101 °C
Density and specific gravity:	1.21 g/mL (liquid under pressure at 25 °C)
Vapor pressure:	96 psi at 25 °C
Flash point and flammability:	Nonflammable
Solubility:	0.15% in water; soluble in ether
Octanol and water partition coefficient:	P_{ow} = 1.06
Conversion factors:	1 mg/L = 238 ppm; 1ppm = 4.2 mg/m^3

respectively (Rissolo and Zapp 1967; Silber and Kennedy 1979a). Also in rats, a 15-min lethal concentration for 50% of the test animals (LC_{50}) was reported to be 800,000 ppm and a 4-hr LC_{50} was reported to be 500,000 ppm (Collins 1984). Clinical signs of toxicity included lethargy, labored and rapid respiration, foaming at the nose, tearing, salivation, convulsions, and death. For surviving animals, the effects were reversible. HFC-134a was not lethal to dogs exposed at concentrations of 700,000-800,000 ppm for 3-5 hr (Shulman and Sadove 1967).

Deep narcosis occurred in dogs, cats, and monkeys exposed via inhalation at concentrations of 500,000 ppm within approximately 1 min; the recovery period was approximately 2 min (Shulman and Sadove 1967). In a review of preclinical toxicology studies, Alexander and Libretto (1995) reported no deaths or treatment-related effects on clinical signs, body weight, food and water consumption, or postmortem findings in rats and mice exposed via inhalation at a concentration of 810,000 ppm with oxygen supplementation for 1 hr. Male and female mice exposed to HFC-134a without oxygen supplementation at 150,000 ppm for 1 hr showed respiratory effects, and the female mice were comatose after 15 min exposure (Alexander and Libretto 1995). In that study, there were significant decreases in tidal volume at 74,000 ppm, slight decreases in respiratory rate at 90,500 ppm, and marked reductions in minute volume at 150,000 ppm. In the same study, rats exposed at concentrations of 47,000 ppm had significantly reduced respiratory rates. Dogs exposed via inhalation to HFC-134a at concentrations of 40,000 and 80,000 ppm for 1 hr did not show treatment-related clinical signs. At 160,000 ppm, three of four dogs showed salivation, head shaking, and struggling; at 320,000 ppm the effects were more severe.

HFC-134a was found to be a weak cardiac sensitizer when tested in an epinephrine challenge in dogs (Mullin and Hartgrove 1979). Cardiac arrhythmias were observed at concentrations of 75,000 ppm and above. No effects were observed at 50,000 ppm. In another study, HFC-134a induced cardiac sensitization at concentrations of 80,000 ppm and above, and no effects were observed at 40,000 ppm (Hardy et al. 1991).

Repeated-Dose Studies

No adverse health effects in humans from repeated exposure to HFC-134a have been reported.

Subacute, subchronic, and chronic studies have been conducted in experimental animals to test the toxicity of HFC-134a (Kennedy 1979; Riley et al. 1979; Silber and Kennedy 1979b; Hext and Parr-Dobrzanski 1993; Hext 1989; Alexander and Libretto 1995; Collins et al. 1995). Those studies are summarized in Table A-4.

The NOAEL for subacute exposure in rats ranged from 10,000 to 100,000 ppm (Kennedy 1979; Riley et al. 1979; Silber and Kennedy 1979b). Subchronic exposure using 1 hr/d snout-only exposure for rats and mice resulted in no effects at 50,000 ppm (Alexander and Libretto 1995); whole-body exposure of rats for 6 hr/d to 50,000 ppm also had no effect (Hext 1989; Collins et al. 1995). There were no effects in dogs when 120,000 ppm was administered by face mask for 1 hr/d (Alexander and Libretto 1995). In rats exposed chronically to HFC-134a by

TABLE A-4 Repeated-Dose Studies on the Toxicity of HFC-134a

Study Type	Species	Exposure Concentration, Duration, Route	Observed Effect	NOAEL (ppm)	LOAEL (ppm)	Reference
Subacute	Rat	10,000, 50,000, 100,000 ppm; 6 hr/d, 5 d/wk for 14 or 28 d; inhalation	Pathological changes in lung (focal interstitial pneumonitis) observed in rats exposed at 50,000 and 100,000 ppm; some changes in organ weight observed, but not related to histological changes	10,000		Silber and Kennedy 1979b
Subacute	Rat	1,000, 10,000, 50,000 ppm; 6 hr/d, 20 times in a 28-d period; inhalation	Changes in kidney and gonad weight in male rats exposed at 50,000 ppm, changes in liver weight in male rats exposed at 10,000 and 50,000 ppm. No pathological changes found. Reduced organ weights not considered of toxicological significance.	50,000		Riley et al. 1979
Subchronic	Rat	2,000, 10,000, 50,000 ppm; 6 hr/d, 5 day/wk for 13 wk with and without 4-wk recovery period; inhalation (whole body)	No treatment-related effects	50,000		Hext 1989

Subchronic	Mouse	10,000, 25,000, 50,000 ppm; 1 hr/d, 7 d/wk for 13 wk; inhalation (snout-only)	No clinical signs related to treatment, no effect on body weight gain	50,000	Alexander and Libretto 1995
Subchronic	Rat	2,500, 10,000, 50,000 ppm; 1 hr/d, 7 d/wk for 50 wk; inhalation (snout-only)	No clinical signs related to treatment, no effect on body weight gain	50,000	Alexander and Libretto 1995
Subchronic	Dog	0.225, 0.75, 2.25 g; 3, 10, or 30 metered doses, twice daily for 1 yr (administered by oropharyngeal tube)	No clinical treatment-related effects		Alexander and Libretto 1995
Subchronic	Dog	120,000 ppm; 1 hr/d for 1 yr; inhalation (face mask)	No clinical signs, changes in body weight, or changes in postmortem findings related to treatment. Minor effects included trembling, salivation, vomiting, were related to treatment-associated anxiety.	120,000	Alexander and Libretto 1995
Subchronic	Rat	2,000, 10,000, 50,000 ppm; 6 hr/d, 5 d/wk for 13 wk; inhalation (whole body)	No evidence of toxicity or compound-related effects at any exposure	50,000	Collins et al. 1995

TABLE A-4 (*Continued*)

Study Type	Species	Exposure Concentration, Duration, Route	Observed Effect	NOAEL (ppm)	LOAEL (ppm)	Reference
Chronic	Rat	2,500, 10,000, 50,000 ppm; 6 hr/d, 5 d/wk for 2 yr; inhalation (whole body)	Survival over 2-yr period not affected by treatment. No changes in clinical condition, clinical chemistry, body weight, food consumption in any exposure group. Histological examination showed statistically significant increase in the incidence of Leydig (interstitial) cell hyperplasia and Leydig cell adenoma in animals exposed to 50,000 ppm	10,000	50,000	Collins et al. 1995; Hext and Parr-Dobrzanski 1993
Chronic	Rat	2,500, 10,000, 50,000 ppm; 1 hr/d, 7 d/wk for 108 wk; inhalation (snout only)	No clinical signs related to treatment, no effect on body weight gain	50,000		Alexander and Libretto 1995

NOAEL, no-observed-adverse-effect level; LOAEL, lowest-observed-adverse-effect level; ppm, parts per million.

inhalation using snout-only exposure for 1 hr/d, 7d/wk there were no effects at 50,000 ppm (Alexander and Libretto 1995); with whole-body exposure for 6 hr/d, 5 d/wk (Hext and Parr-Dobrzanski 1993; Collins et al. 1995), an increase in Leydig cell hyperplasia and adenomas was seen at 50,000 ppm with a NOAEL of 10,000 ppm.

Genetic Toxicity Studies

There is no evidence to suggest that HFC-134a induces either genetic or chromosomal mutations, and therefore, there is no reason to suspect that HFC-134a exposure would induce heritable effects in humans. HFC-134a is reported to be nonmutagenic when tested in the Ames assay (Litton Bionetics 1976; Callander and Priestly 1990; Collins et al. 1995) or in the microbial mutagenicity assay in *Escherichia coli* (Alexander and Libretto 1995). It does not alter DNA synthesis in rat hepatocytes (Trueman 1990; Collins et al. 1995), induce chromosomal aberrations in mouse lymphoma L51787 cells in the presence or absence of microsomal-induced liver homogenates (Alexander and Libretto 1995), human lymphocytes or Chinese hamster lung cells (Mackay 1990; Collins et al. 1995), or alter micronucleus formation in the femoral bone marrow of exposed mice (Muller and Hoffmann 1989; Collins et al. 1995). HFC-134a also appears to be nonmutagenic to male mice exposed at 1,000, 10,000, or 50,000 ppm for 6 hr/d for 5 d via inhalation when tested in a dominant lethal assay (Hodge et al. 1979a). The results of a study of chromosomal aberrations in rat bone marrow cells were inconclusive (Anderson and Richardson 1979).

Carcinogenicity Studies

Rats were exposed, whole body, to HFC-134a via inhalation at concentrations of 2,500, 10,000, and 50,000 ppm for 6 hr/d, 5d/wk for up to 104 wk (Hext and Parr-Dobrzanski 1993; Collins et al. 1995). At 50,000 ppm there was an increase in testicular weight, Leydig cell hyperplasia, and Leydig cell tumors. Such tumors are common in rats and are induced by a variety of chemicals. Because HFC-134a does not demonstrate mutagenic activity, the increased incidence of Leydig cell tumors is attributable to a nongenotoxic mechanism. A very low incidence of Leydig cell tumors in humans has been reported (Mostofi and Price 1973), and the relevance of extrapolating the findings from rats to humans has been questioned. However, Clegg et al. (1997), in a thorough evaluation of Leydig cell hyperplasia and adenomas, indicated that the incidence in humans is uncertain and that, as a default when the mode of induction is unknown, agents that induced both hyperplasia and adenomas should be considered relevant and of concern for progression to carcinogenesis in humans. Several other HFCs also have been shown to induce Leydig cell hyperplasia and adenomas (summarized in Clegg et al. 1997). When this is the only or the primary effect of an agent and there is no mutagenic activity, the dose-response relationship is assumed to be nonlinear.

In another study, rats were exposed to HFC-134a at 300 mg/kg of body weight by gavage 5 d/wk for 1 year (Longstaff et al. 1984). No carcinogenicity was observed in this investigation. However, only one concentration was used, and it is possible that the route of administration and the dose of the compound used were not capable of detecting carcinogens of low potency. Furthermore, because the onset of Leydig cell adenomas is usually seen in aged animals, chronic studies will be more useful for detecting these effects and predicting their occurrence.

Other Toxicity

HFC-134a was shown to cause slight skin irritation in rabbits, perhaps because of local freezing (Mercier 1989). In that study, 0.5 mL of liquified HFC-134a was applied to scarified and intact skin areas of rabbits and the exposed site was covered for up to 24 hr.

HFC-134a also was shown to produce slight eye irritation in rabbits (Mercier 1990a). The chemical was administered as a gas, sprayed for either 5 or 15 seconds (sec) from a distance of 10 centimeters.

HFC-134a did not produce skin sensitization in one study conducted in guinea pigs (Mercier 1990b). The animals received a single intradermal injection of Freund's complete adjuvant followed by seven consecutive (occlusive) epicutaneous administrations of liquified HFC-134a. The challenge administration was performed after 12 d without treatment by occlusive epicutaneous treatment with liquified HFC-134a.

Pharmacokinetics

Two pharmacokinetics studies have been conducted in human volunteers with exposure to HFC-134a. Vinegar et al. (1997) reported on the exposure of two male volunteers to 0.4% (4,000 ppm) HFC-134a via inhalation. The first subject lost consciousness after 4.5 min and exhibited a dramatic increase in blood concentration, which reached 1.29 mg/L by 2.5 min, and decrease to zero in pulse rate and blood pressure. The subject was revived and pulse and blood pressure returned to normal after 1 hr. However, dizziness and balance problems persisted after 6 wk. The second subject showed a rapid rise in blood pressure and pulse after 10.5 min, by which time the blood concentration had reached 0.7 mg/L; he was removed from the exposure and his vital signs returned to normal after 30 sec. The same subject was exposed again after about 1 hr to 0.2% HFC-134a and began having problems after 2.5 min. His blood concentration was 0.16 mg/L at the beginning of exposure and reached 0.38 mg/L by 2.5 min. Most symptoms were gone by the next day, but dizziness and balance problems persisted for 6 wk and he reported persistent ringing in the ears.

The second study (Emmen and Hoogendijk, 1999) involved eight volunteers (four males and four females). This study used whole-body inhalation exposure to CFC-12 or HFC-134a on eight occasions. The exposures to HFC-134a were at 1,000, 2,000, 4,000, or 8,000 ppm for 1 hr. No treatment-related effects of inhalation exposure to HFC-134a were noted on echocardiogram, pulse rate, systolic and diastolic blood pressure, or lung function compared with air control and CFC-12 reference conditions. The maximum concentrations reached 5.95-7.22 µg/mL after exposure to 8000 ppm. The half time $(t_{1/2})$ for distribution was 8.34-9.44 min, and for elimination it was 38.29-44.45 min.

Absorption of fluorocarbons and bromofluorocarbons via inhalation in experimental animals is rapid; the maximal blood concentrations of the substances develop within 5 min and equilibrium is achieved within the next 15 min of exposure (Azar et al. 1973; Trochimowicz et al. 1974; Mullin et al. 1979). Blood concentrations do not increase further with increasing durations of exposure for a given concentration of these substances. In a fertility study (Alexander et al. 1996), blood samples were taken from P (parental generation) male rats after 15 wk exposure, and from P females after 3 wk premating, 3 wk gestation, and 2 wk postpartum exposure. P females from a peri- and postnatal study were sampled after one exposure on gestation day 17, or after repeated exposures in the second week postpartum. The data showed rapid absorption into blood of HFC-134a, increasing concentration with increasing exposure, and rapid excretion with no accumulation on repeated dosing. The mean half-lives ranged from 5.8 to 7 min.

Toxic effects observed in animals following oral and inhalation exposure to HFC-134a indicate it is absorbed by the lungs and gastrointestinal tract (Salmon et al. 1980). Studies conducted in rats exposed to high concentrations of HFC-134a, either orally or via inhalation, indicate it is rapidly excreted, mostly as the unchanged parent compound (Salmon et al. 1980). Analysis of the urine, feces, and expired air of rats exposed to HFC-134a at 10,000 ppm (1.0%) for 1 hr showed that only 0.34%-0.40% was metabolized (Ellis et al. 1993). The study by Salmon et al. (1980) found that some HFC-134a is retained in the liver and that relatively large amounts are retained in the adrenal gland; the study by Ellis et al. (1993) did not report any evidence for specific accumulation in any organ or tissue, including fat.

Studies in rat liver microsomes show that HFC-134a is oxidized by the cytochrome P450 system; this implies that cytochrome P450-containing tissues, such as nasal mucosa, liver, and lungs might convert HFC-134a to trifluoracetic acid, a toxic metabolite (Olson et al. 1990). Another study on the metabolism of HFC-134a in isolated rat hepatocytes found that the chemical undergoes limited metabolism as measured by the release of inorganic fluoride (Reidy et al. 1990). Microsomal metabolism was inhibited by carbon monoxide, was decreased in the presence of low oxygen concentration, and was increased in the presence of hepatic microsomes isolated from Arochlor-treated rats. These results indicate that HFC-134a undergoes a cytochrome P450-catalyzed defluorination reaction.

Reproductive and Developmental Toxicity Data

Human Studies

No data were found on the reproductive and developmental effects of HFC-134a in humans.

Experimental Animal Studies

Reproductive Toxicity

Three reproductive toxicity studies testing HFC-134a have been conducted in rats and mice (Table A-5). Additionally, data from several repeated-dose studies included information on reproductive effects (see Table A-4).

Hodge et al. (1979a) exposed male CD-1 mice (40 per group) in a dominant lethal study design to HFC-134a at 0, 1,000, 10,000 or 50,000 ppm by inhalation for 6 hr/d for 5 consecutive days, then mated them to unexposed females (two females for 4 d/wk) for 8 wk. Fifteen days after the initial date of pairing, the females were killed and their uterine contents examined. Females were not examined for vaginal plugs so there was likely some variability in gestational age when females were killed. The only dose-related effects observed were an increase in early deaths (as a percentage of implants) in the 50,000-ppm group at wk 4 and 8. This could have been because of a low incidence in the controls at these times; the incidence in the HFC-134a-exposed groups was within the control range across all weeks and was unlikely to have been a significant treatment effect.

In a two-generation study by Alexander et al. (1996), 30 male and female rats per group were exposed by snout-only exposure to HFC-134a at 0, 2,500, 10,000 or 50,000 ppm (1 hr/d) for 10 wk (males) or 3 wk (females) before pairing (P generation), the males continued to be exposed for a total of 18 wk, and the females continued through pregnancy and lactation. Estrous cycles were monitored for 14 d before pairing. F_1 (first filial) generation fetuses were examined on day 20 of gestation in 14 females per group; the rest were allowed to litter and nurse their young. Twelve F_1 males and females per group were selected at weaning and mated at approximately 70 d of age. Survival and development of the F_2 (second filial) generation progeny were monitored for 21 d postpartum. One F_2 animal of each sex (eight litters per group) was retained to sexual maturity. Physical and reflex development were evaluated in F_1 and F_2 offspring, as were locomotor coordination, activity, and learning, memory, and reversal. A slight but statistically significant decrease in body weight gain was seen in P males exposed to 10,000 and 50,000 ppm after 2 wk exposure; cumula-

TABLE A-5 Studies Testing Reproductive and Developmental Toxicity of HFC-134a

Study Type	Species	Exposure Concentration, Duration, Route	Observed Effect	NOAEL (ppm)	LOAEL (ppm)	Reference
Dominant lethal study	Mice (male)	1,000, 10,000, 50,000 ppm; 6 hr/d for 5 d; inhalation (whole body). After exposure, mated with 2 females each week for 8 wk; females killed 15 d after pairing	No significant maternal or developmental toxicity	≥50,000		Hodge et al. 1979a
Two-generation study (paternal exposure only)	Rat (male and female)	2,500, 10,000, 50,000 ppm; 1 hr/d; males treated from 10 wk before mating (period of gametogenesis) through mating (18 wk total exposure), females treated from 3 wk before mating (gametogenesis) through mating, pregnancy, and lactation; inhalation (snout-only). F₁ offspring exposed via nursing, not exposed after weaning, F₂ offspring not exposed	No adverse effects on reproductive performance of treated animals. Slight reduction in body weight gain of males exposed at 10,000 and 50,000 ppm; significant reduction of male body weight gain at 50,000 ppm after 5 and 10 wk exposure. No effect on offspring growth survival or body weight. F₂ offspring showed minor delays (~1/2 to 1 d) in physical, reflex development; not considered related to exposure	Adult: 10,000 Offspring: ≥50,000	Adult: 50,000	Alexander et al. 1996

				NOAEL	LOAEL	Reference
Testicular endocrine function	Rat (male)	10,000, 30,000, 100,000 ppm; 6 hr/d; animals treated through gametogenesis (11 wk) and mating and postmating period (18 weeks total); inhalation, first 9 wk snout-only, final 9 wk whole-body	Levels of luteinizing hormone measured. No difference between controls and treated groups. Animals exposed at 100,000 ppm showed slight (not statistically significant) increase in testosterone secretion and biosynthesis and a concomitant rise in progesterone secretion when the testis was incubated with human chorionic gonadotrophin, without any qualitative change in androgen biosynthesis	30,000	100,000	Barton et al. 1994
Prenatal developmental toxicity	Rat	30,000, 100,000, 300,000 ppm; 6 hr/d from d 5-14 of gestation; inhalation (whole body)	At 300,000 ppm, significant decrease in fetal weight, significant increase in skeletal variations in fetuses. Maternal toxicity observed at 100,000, 300,000 ppm. No developmental effects in fetuses exposed at 30,000 or 100,000 ppm	Maternal: 30,000 Fetal: 100,000	Maternal: 100,000 Fetal: 300,000	Lu and Staples 1981

TABLE A-5 *(Continued)*

Study Type	Species	Exposure Concentration, Duration, and Route	Observed Effect	NOAEL (ppm)	LOAEL (ppm)	Reference
Prenatal developmental toxicity	Rat	1,000, 10,000, 50,000 ppm; 6 hr/d from days 6-15 of gestation; inhalation (whole body)	At 50,000 ppm, significantly reduced fetal weight, skeletal ossification. No significant maternal toxicity	Maternal: 50,000 Fetal: 10,000	Fetal: 50,000	Hodge et al. 1979b
Prenatal developmental toxicity	Rabbit	Pilot study: 5,000, 20,000, 50,000 ppm (6-9 pregnant animals per group), 6 hr/d days 6-18 gestation; inhalation (whole body) Main study: 2,500, 10,000, 40,000 ppm (18-24 pregnant animals per group, 6 hr/d from days 6-18 gestation; inhalation (whole body)	Pilot study: Slight decrease in maternal body weight gain at 50,000 ppm. Decreased number of implantations in all groups, significant at 20,000 and 50,000 ppm; decreased number of live fetuses, gravid uterine weight, litter weight, and increased fetal weight at all exposure levels Main study: At 10,000 and 40,000 ppm, maternal body weight and food consumption reduced; effects at 10,000 ppm within historical control range. No treatment-related effects on implants, litter size, fetal weight, or external, visceral, skeletal defects at any exposure level	Maternal: 10,000 Fetal: ≥40,000	Maternal: 40,000	Wickrama-ratne 1989a,b (same as study reported by Collins et al. 1995)

| Peri- and postnatal | Rat | 1,800, 9,900, 64,400 ppm; 1 hr/d; days 17–20 pregnancy and 1–21 postpartum; inhalation (snout-only) | No maternal or developmental effects observed except for a delay in F_1 age at pinna detachment, eye opening and startle response at 64,400 ppm (~1/2-d delay). May be related to exposure | Maternal: 64,400

Fetal: 9,900 | Fetal: 64,400 | Alexander et al. 1996 |

NOAEL, no-observed-adverse-effect level; LOAEL, lowest-observed-adverse-effect level; ppm, parts per million.

tive body weight gain in males exposed to 50,000 ppm over 5 wk or 10 wk was significantly reduced. There were no effects on body weight or weight gain in P females, or in F_1 or F_2 offspring. There appeared to be a slight increase in skeletal defects in F_1 fetuses, but no significant change was reported, and some might have been cases of decreased ossification (this was unclear from the reported data). In F_2 offspring, there was a slight but statistically significant increase in the age at pinna detachment (exposed at 10,000 ppm), startle response (exposed at 10,000 and 50,000 ppm), and air righting (exposed at 2,500 and 10,000 ppm); these changes were not clearly dose related, and on average they represented a ½- to 1-d delay. Because the F_2 offspring were never exposed directly or indirectly, there was no change in body weight at birth or weaning, and no changes were seen in the F_1 offspring on these same measures; the changes detected in F_2 animals were not considered treatment-related.

As a follow-up to studies showing Leydig cell hyperplasia, Barton et al. (1994) exposed 25 male Sprague-Dawley rats per group to HFC-134a at 0, 10,000, 30,000, or 100,000 ppm for 6 hr/d for a total of 18 wk (11 wk before mating and 7 wk during and after mating). Animals were exposed snout-only for the first 9 wk to reduce the amount of material used; thereafter, whole-body exposure was used. In 10 males per group, luteinizing hormone concentrations were assessed after 16 wk, and again at 17 wk after stimulation with luteinizing hormone releasing hormone (LHRH). In another 10 males per group, at necropsy, the left testis was decapsulated, then incubated with human chorionic gonadotropin to assess androgen release; the right testis was examined histologically. High basal concentrations of luteinizing hormone were seen in all groups including controls, but there was no difference between controls and treated groups in luteinizing hormone levels before or after LHRH stimulation. At 100,000 ppm, there was no statistically significant increase in testosterone secretion and biosynthesis, but an increase in progesterone was observed. The increase in progesterone was consistent with increased Leydig cell function at this exposure level.

Developmental Toxicity Studies

Four developmental toxicity studies testing HFC-134a have been conducted in rats and rabbits (Table A-5).

Lu and Staples (1981) exposed female Sprague-Dawley rats at concentrations of 0, 30,000, 100,000, or 500,000 ppm HFC-134a for 6 hr/d, days 5-14 of gestation (11 pregnant control animals, 6 in each exposed group). All gestational ages are converted to correspond to the day of insemination as gestational day 0. Animals were killed on gestational day 20 and uterine contents were examined. Dams exposed to 100,000 or 300,000 ppm showed a reduced or absent response to sound, respectively, demonstrating the anesthetic action of HFC-134a. Animals exposed to 300,000 ppm consumed significantly less food and gained significantly less weight than did controls. There was a significant decrease in fetal weight at 300,000 ppm, and a significant increase in the incidence of skeletal variations, many of which were related to reduced ossification.

Hodge et al. (1979b) exposed female rats to concentrations of 0, 1,000, 10,000, or 50,000 ppm HFC-134a for 6 hr/d on gestation days 6-15 and killed them on gestation day 21 for examination of uterine contents (23-29 pregnant animals per group). There were no treatment-related effects on maternal animals except for acute pulmonary irritation that increased in severity and incidence with exposure concentration. There was no effect on the number of implants, litter size, or litter weight. At 50,000 ppm, fetal weight was slightly but significantly decreased and there was an increased incidence of skeletal variations, primarily reduced ossification of cervical vertebrae, sternebrae, and digits. An increase in the incidence of abnormal sternebrae also was reported in the 50,000 ppm group, but these effects (bipartite or mis-aligned sternebrae) are often seen in controls and are likely related to reduced ossification observed in the same groups. All fetal effects reported in this study at the highest exposures can be accounted for by the fact that litter size was greater in this group than in any other, including controls. Reduced fetal weight often is associated with increased litter size, and reduced ossification of skeletal elements often accompanies reduced fetal weight.

Wickramaratne et al. (1989a,b) conducted a prenatal developmental toxicity study in rabbits in two phases. In the pilot study (Wickramaratne et al. 1989a), groups of artificially inseminated rabbits were exposed to 5,000, 20,000 or 50,000 ppm (six to nine pregnant animals per group) for 6 hr/d, on gestation days 6-18, then killed on gestation day 29. Fetuses were examined only for external anomalies and cleft palate. Two animals aborted late in pregnancy, one in the

5,000 ppm group and one in the 50,000 ppm group. Maternal body weight loss was observed during the early dosing period in the 50,000 ppm group. The number of implantations was reduced in the 20,000 and 50,000 ppm treated groups compared with controls, with consequent reduction in litter size, gravid uterine weight, and litter weight, but increased fetus weight due to reduced litter size.

In the main study (Wickramaratne et al. 1989b; same study reported in Collins et al. 1995), pregnant rabbits (18-24 per dose group) were exposed 6 hr/d to 2,500, 10,000, and 40,000 ppm gestation days 6-18. Observations were the same as in the pilot study, but fetuses were examined for external, visceral, and skeletal defects, and a single section was made through the head to examine the brain macroscopically. One control animal aborted late in pregnancy. Maternal body weight and food consumption were signficantly reduced at 10,000 and 40,000 ppm, although the effects at 2,500 ppm were within the historical control range. There was no effect on implantation number, litter size, gravid uterine weight, litter weight, or individual fetus weight. The incidence of major and minor defects either did not appear to be dose related or was within historical control ranges. The NOAEL was considered to be 10,000 ppm, based on maternal toxicity, and the NOAEL for developmental toxicity was ≥40,000 ppm.

In a peri- and postnatal study design, Alexander et al. (1996) exposed rats by snout only to HFC-134a at 0, 1,800, 9,900, or 64,400 ppm (1 hr/d). Pregnant females were exposed on gestation day 17-20, then from postnatal day 1-21. F_1 pups were weaned at postnatal day 21, and one male and female per litter (20 litters per group) were selected and raised to maturity. F_1 animals were mated at approximately 84 d of age, killed on gestation day 20 , and the uterine contents were examined. F_1 offspring were examined for physical and reflex development as well as locomotor coordination, activity, learning, memory, and reversal. There were no effects on any parameter measured except for a statistically significant delay in the age at pinna detachment, eye opening, and startle response in F_1 offspring at 64,400 ppm (approximately half-day delay). Although there was no effect on body weight, these animals were exposed indirectly via the milk during the period when most of these measurements were made. Exposure concentrations and blood concentrations of the dams were somewhat higher in the peri- and postnatal study than in the fertility study reported at the same time. These minor but statistically significant effects could be the result of the exposure to HFC-134a.

Integration of Toxicity and Exposure Information

Interpretation of Toxicity Data

General Toxicity and Pharmacokinetics

Data on the toxicity of HFC-134a indicate that it is nontoxic in most circumstances. Most of the changes reported have been at high concentrations, which also induce narcosis. There is no evidence of genetic toxicity for HFC-134a. The primary effect reported is the induction of Leydig cell hyperplasia and adenomas in male rats exposed at 50,000 ppm for 6 h/d, 5 d/wk over a 2-yr period (Collins et al. 1995; Hext and Parr-Dobrzanski 1993). The NOAEL of 10,000 ppm from these data was the basis of the EPA's RfC.

It should be noted that in a two-generation study (Alexander et al. 1996), the NOAEL for adult toxicity was 10,000 ppm, based on significant changes in body weight in male rats exposed snout-only at a concentration of 50,000 ppm for 1 h/d, 5 d/wk for 18 wk. The difference in results might be attributed to the snout-only exposure protocol, which might have caused stress to the rats and affected their body weight in the Alexander et al. (1996) study.

Data in two human volunteers from one study (Vinegar et al. 1997) indicate severe effects on vital signs after brief exposures at concentrations of 2,000-4,000 ppm HFC-134a by inhalation. However, in a study by Emmen and Hoogendijk (1999), human volunteers exposed whole body at concentrations as high as 8000 ppm for 1 hr did not show any effects of HFC-134a. Absorption of HFC-134a was very rapid and maximal concentrations were achieved within 15-30 min. Repeated exposures in animal studies do not result in accumulation because the half-life is so short. Given the rapid absorption and excretion in rats and humans, the kinetics appear to be similar between the species.

Reproductive and Developmental Toxicity

Interpretation of the experimental animal data described above is complicated by the fact that two exposure regimens were used: 1 hr/d snout-only exposure, to mimic MDI exposures, and 6 hr/d whole-body exposure, to mimic environmental exposures.

Developmental Toxicity. No data were found from studies of developmental toxicity of HFC-134a in humans. The experimental animal data are *sufficient* to conclude that HFC-134a does not cause prenatal developmental toxicity when pregnant animals were exposed (whole body) for 6 hr/d during major organogenesis to HFC-134a concentrations below those associated with narcosis (<30,000 ppm). The NOAEL for rat maternal toxicity was 30,000 ppm, whereas the NOAEL for rat prenatal toxicity was 50,000 ppm (Hodge et al. 1979a; Lu and Staples 1981). The NOAEL for rabbit maternal toxicity was 10,000 ppm, whereas the NOAEL for rabbit prenatal toxicity was greater than 40,000 ppm (Wickramaratne 1989a,b; Collins et al. 1995). The data are *insufficient* to determine the extent to which HFC-134a causes postnatal developmental effects, as the only study addressing this issue used 1 hr/d snout-only exposures for 10 wk before mating, throughout gestation to day 20, and from postnatal days 1-21 (fertility study), or from gestation day 17-20 and postnatal days 1-21 (peri- and postnatal study); exposure to pups was not continued beyond post-natal day 21 (Alexander et al. 1996). In these studies, the adult NOAEL was 10,000 ppm (LOAEL, 50,000 ppm) in the fertility study and greater than 64,400 ppm in the peri- and postnatal study. The developmental NOAEL was greater than 50,000 ppm in the fertility study and 9,900 ppm (LOAEL= 64,400 ppm) in the peri- and postnatal study. The effects at 64,400 ppm in the peri- and postnatal study were minimal and might not be related to exposure, so 50,000 ppm is assumed to be the NOAEL for postnatal effects for snout-only exposure for 1 hr/d. The data are *insufficient* to determine what the extent and types of effects would be with continued exposure to pups after weaning and into the F_2 generation. The experimental animal data are *assumed relevant* to humans.

Reproductive Toxicity. No data were located from studies of the reproductive toxicity of HFC-134a in humans. The animal data were *sufficient* to show that HFC-134a exposures similar to those metered-dose inhalers did not affect fertility and sexual function. Although there was an effect of exposure to 50,000 ppm on male body weight gain in the fertility study mentioned above (Alexander et al. 1996), fertility and reproductive function were not affected.

The data were *sufficient* to show that whole-body exposure to HFC-134a for 6 hr/d for 2 yr could affect the testis. Although a dominant lethal study showed no effects at concentrations as high as 50,000 ppm (Hodge et al. 1979a), chronic exposure of rats at 50,000 ppm HFC-134a resulted in a statistically significant increase in the incidence of Leydig cell hyperplasia and adenoma (Hext and Parr-Dobrzanski 1993; Collins et al. 1995). The NOAEL was 10,000 ppm for a 6 hr/d exposure. In a follow-up study to Hext and Parr-Dobrzanski (1993), Barton et al. (1994) showed that HFC-134a (6 hr/d, 18 wk) increased the synthesis and release of testosterone and progesterone from testes incubated with human chorionic gonadotropin but did not alter the qualitative aspects of androgen biosynthesis. The changes in testosterone and progesterone secretion were consistent with the increased Leydig cell activity in the chronic study. Given the significant increase in Leydig cell hyperplasia and adenoma in rats, EPA considered these effects adverse and based the RfC for HFC-134a on this effect.

The experimental animal data are *assumed relevant* to humans, because the data are inadequate to show that the effects are irrelevant.

Default Assumptions

The data on Leydig cell hyperplasia and adenoma are assumed to be relevant to humans. Likewise, the lack of developmental toxicity is also assumed to be relevant to humans. The pharmacokinetics in humans and animals are very similar, as is the lack of toxicity from acute exposures, notwithstanding the data from the two subjects reported by Vinegar et al. (1997). Because of the similarities in human and animal pharmacokinetics, this portion of the interspecies uncertainty factor can be reduced from 10 to 3. The intraspecies uncertainty factor of 10 should be retained because of the unusual findings in the Vinegar et al. (1997) study. A database deficiency factor of 10 should be applied due to the lack of a 6 hr/d exposure in a multigeneration study with exposure continuing throughout the two generations, as well as lack of a developmental neurotoxicity study with a 6 hr/d exposure. It appears that there is little or no effect of HFC-134a in any of the studies reviewed except at doses that induce narcosis, but additional data with longer exposure durations and additional endpoints are needed to confirm this observation.

Quantitative Evaluation

In the two-generation study by Alexander et al. (1996), no adverse reproductive or developmental effects were observed in rats exposed at a concentration of 50,000 ppm for 1 hr/d during gametogenesis and mating (males), or gametogenesis, mating, pregnancy, and lactation (females). Since the effects at 64,400 ppm (exposed for 1 hr/d) in the peri- and postnatal study (Alexander et al. 1996) were minimal, and the next lower dose was 9,900 ppm, the subcommittee has identified 50,000 ppm as the NOAEL for reproductive and developmental effects for less than a chronic exposure. The NOAEL of 50,000 ppm for a 1 hr/d exposure converted to a 6 hr/d exposure is 8,300 ppm. The UEL for reproductive and developmental effects is then calculated using the adjusted NOAEL of 8,300 ppm divided by an aggregate uncertainty factor of 300 (3 for interspecies extrapolation, 10 for intraspecies differences, and 10 for deficiencies in the database). Thus, the UEL for reproductive and developmental toxicity is

$$\frac{8{,}300 \text{ ppm}}{300} = 28 \text{ ppm for a less than chronic exposure.}$$

For chronic exposure, the subcommittee chose the NOAEL of 10,000 ppm for a 6-hr/d, 2-yr exposure based on a significant increase in the incidence of Leydig cell hyperplasia in treated rats (Collins et al. 1995; Hext and Parr-Dobrzanski 1993). Applying an aggregate uncertainty factor of 300 (3 for interspecies extrapolation, 10 for intraspecies differences, and 10 for deficiencies in the database), the UEL for reproductive and developmental toxicity is

$$\frac{10{,}000 \text{ ppm}}{300} = 33.3 \text{ ppm } (140 \text{ mg/m}^3) \text{ for a chronic exposure.}$$

The chronic exposure UEL is higher than the EPA's chronic RfC of 80 mg/m^3, which is based on the same study and endpoints (see above). There are several differences in the way the UEL was derived compared with the RfC. To calculate the chronic RfC, EPA derived a BMC$_{10}$ of 11,030 ppm and adjusted it from a 6-hr/d, 5-d/wk exposure to a continuous exposure by multiplying by 6/24 hr and 5/7 d. A total uncertainty factor of 100 was applied to the BMC_{10}: 3 for interspecies extrapolation, 10 to protect sensitive individuals, and 3 for a database deficiency of a two-generation study. The NOAEL of 10,000 ppm was used as the basis for the chronic UEL and the UEL was not adjusted because it was developed for an occupational exposure scenario. A total uncertainty factor of 300 was applied to the NOAEL: 3 for interspecies extrapolation, 10 to protect sensitive individuals, and 10 for database deficiencies due to the lack of both a two-generation study and a developmental neurotoxicity study. The latter concern was raised by the Alexander et al. (1996) peri- and postnatal study that suggested the possibility of developmental neurotoxicity in offspring of animals exposed for 1 hr/d, 5 d/wk during late gestation and lactation. Had the BMC_{10} derived by EPA been used in the calculation, the final value would have been slightly higher (i.e., 37 ppm or 155 mg/m^3).

Critical Data Needs

A two-generation reproduction study is needed with at least 6 hr/d exposure continuing to pups after weaning and into the F_2 generation to determine the effects of long-term exposures. Developmental neurotoxicity endpoints should be included in this study based on the types of effects seen in the peri- and postnatal study with the snout-only 1 hr/d exposure.

Summary

HFCs, including HFC-134a, have been developed as an alternative to CFCs, which are known to contribute to the breakdown of ozone to oxygen in the stratosphere. HFCs do not contribute to the destruction of stratospheric ozone, but some HFCs have global warming potential. They primarily serve as replacements for CFCs in refrigeration equipment and mobile air conditioning; they also have pharmaceutical applications (e.g., as propellants for metered-dose inhalers used to treat asthma).

Human Exposure

Human exposure to HFC-134a occurs via inhalation from accidental leaks of air conditioning units and refrigerators, from spills or industrial use, and from use of metered-dose inhalers such as those that deliver medication for the treatment of asthma.

Toxicology

Developmental Toxicity

There are no human data on the effects of HFC-134a on development. The animal data are sufficient to support a conclusion that exposures to HFC-134a does not cause prenatal developmental toxicity when pregnant animals are exposed for 6 hr/d during major organogenesis to concentrations of HFC-134a below those associated with narcosis (less than 30,000 ppm). HFC-134a also did not cause peri- and postnatal effects in rats with snout-only exposure for 1hr/d during gestation days 17-20 and postnatal days 1-21 at 50,000 ppm. However, the data are insufficient to determine the extent to which HFC-134a caused postnatal developmental effects including endpoints of developmental neurotoxicity, as there are no studies with at least a 6 hr/d exposure throughout two generations. These toxicity findings are assumed to be relevant for the prediction of risk to humans.

Reproductive Toxicity

There are no human data on the effects of HFC-134a on male or female reproduction. The animal data are sufficient to conclude that HFC-134a exposures similar to those used in MDIs will not cause affect sexual function and fertility. There was no effect on dominant lethality after treatment of males and mating with unexposed females. However, data are insufficient to determine the effects of whole-body exposure for 6 hr/d, as a multigeneration study with exposure continuing through two generations is not available. Chronic exposure of rats at concentrations as high as 50,000 ppm resulted in a significant increase in the incidence of Leydig cell hyperplasia. These toxicity findings are assumed to be relevant for the prediction of human risk.

Quantitative Evaluation

Reproductive and Developmental Toxicity

There are no human data from which to develop a quantitative evaluation. Laboratory studies in rats identified a NOAEL of 50,000 ppm for reproductive and developmental effects for a less than chronic exposure. The NOAEL of 50,000 ppm was adjusted to convert from a 1hr/d exposure to a 6hr/d exposure; the adjusted NOAEL is 8,300 ppm. Dividing the adjusted NOAEL by an aggregate uncertainty factor of 300 (3 for interspecies extrapolation, 10 for intraindividual differences, and 10 for an incomplete database), the UEL for reproductive and developmental toxicity for less than a chronic exposure is 28 ppm.

A similar UEL is calculated for chronic exposures. Laboratory studies in rats identified a NOAEL of 10,000 ppm based on a significant increase in the incidence of Leydig cell hyperplasia. Applying an aggregate uncertainty factor of 300 (3 for interspecies extrapolation, 10 for intraindividual differences, and 10 for an incomplete database), the UEL for reproductive and developmental toxicity for a chronic exposure is 33.3 ppm.

Certainty of Judgment and Data Needs

Data on the toxicity and disposition of HFC-134a suggest that it is a compound with little or no reproductive or developmental toxicity except at very high exposure concentrations that induce narcosis. However, data are needed from postnatal evaluations and from a multigeneration study with 6 hr/d exposure. The similarities in pharmacokinetics between humans and laboratory animals provide confidence that the data are relevant for predicting human risk.

Appendix B

Ascertaining Information on the Reproductive and Developmental Toxicity of Agent Exposures

Information on potential reproductive and developmental toxicity of exposures to chemical and physical agents can be obtained from a variety of resources that vary greatly in their approach, coverage, and manner of presentation. Some resources include only bibliographic information; others provide critical evaluations of the available data sets. Some resources emphasize investigations in humans; others focus on experimental animal studies. Some resources take a quantitative approach; others provide only qualitative descriptions of available data. Some resources include only information published in the open literature; others emphasize studies, often unpublished, that have been submitted for regulatory review.

This appendix describes many of the information resources available on reproductive and developmental toxicology that should be useful to the Navy for conducting evaluations. These resources are listed in Boxes B-1 and B-2. The amount of relevant information varies greatly from resource to resource and from agent to agent and for different specific reproductive and developmental outcomes. The following section provides descriptions of several documents and databases that are designed specifically to evaluate reproductive and

Box B-1 Sources of Information Specific to Reproductive and Developmental Toxicity

Detailed Evaluations	Informational Summaries	Bibliographic Resources	Primary Data
California Environmental Protection Agency Hazard Identification Documents for Reproductive and Developmental Toxicity	Reproductive Hazards of Industrial Chemicals: An Evaluation of Animal and Human Data	Developmental and Reproductive Toxicology (DART) and Environmental Teratology Information Center (ETIC) Database	National Toxicology Program Teratology Studies
	REPROTEXT		National Toxicology Program Short-Term Developmental and Reproductive Toxicity Studies
Evaluative Process (Moore et al. 1995 a,b; 1997)	REPROTOX		
	Chemically Induced Birth Defects		National Toxicology Program Continuous Breeding Studies
National Institute of Environmental Health Sciences Center for the Evaluation of Risks to Human Reproduction Documents	Shepard's Catalog of Teratogenic Agents		
	Teratogen Information System		

Box B-2 Additional Sources That May Include Information on Reproductive and Developmental Toxicity

Detailed Evaluations	Informational Summary	Bibliographic Resource	Primary Data
Agency for Toxic Substances and Disease Registry Toxicological Profiles	Chemical Hazards in the Workplace	National Institute for Occupational Safety and Health Registry of Toxic Effects of Chemical Substances	National Toxicology Program Toxicology Report Series
National Institute for Occupational Safety and Health Criteria Documents		MEDLINE	Organization for Economic Cooperation and Developmental Screening Information Data Set Profiles
International Agency for Research on Cancer Monographs on the Evaluation of Carcinogenic Risks to Humans		TOXLINE	
International Programme on Chemical Safety Environmental Health Criteria Documents		PUBMED	
Environmental Protection Agency Integrated Risk Information System			
Environmental Protection Agency National Center for Environmental Assessment Documents			

developmental toxicity data. Generally, except for the bibliographic databases and primary data sources, an expert or committee of experts in reproductive and developmental toxicology has either written or reviewed each document or database. The final section describes additional sources of information that might be useful for assessing agent exposures for potential reproductive and developmental toxicity. Quality control, including the extent of peer review, is noted for each source, as well as the subcommittee's evaluation of the usefulness of the source in identifying exposures that pose a substantial risk of reproductive or developmental toxicity in humans.

No single source of information exists that includes everything needed for a comprehensive evaluation of the reproductive and developmental toxicity of exposures to most agents. It is usually necessary to review several secondary information sources as well as a large amount of primary data to obtain an adequate overall assessment. Even such a thorough review of available information is inadequate in most cases because all of the necessary studies to fully delineate the reproductive and developmental toxicity of an agent have not been conducted. The most complete screening data sets are for pesticides, pharmaceutical agents, food and color additives, and some environmental chemicals (e.g., lead, methylmercury, and polychlorinated biphenyls). The proprietary nature of the data for the first three cases may require the Navy to request information from the appropriate regulatory agencies.

SOURCES OF INFORMATION SPECIFIC TO REPRODUCTIVE AND DEVELOPMENTAL TOXICITY

Detailed Evaluations

California Environmental Protection Agency Hazard Identification Documents on Reproductive and Developmental Toxicity

Description

The Safe Drinking Water and Toxic Enforcement Act of 1986 (Proposition 65, California Health and Safety Code 25249.5 et seq.) requires that the governor cause to be published a list of those chemicals "known to the state" to cause cancer or reproductive (including developmental) toxicity. Substances can be added to that list if they have been classified as reproductive toxicants by an authoritative body (e.g., the World Health Organization, WHO), by the state or federal government, or by a state expert panel, including the California Office of Environmental Health Hazard Assessment Science Advisory Board Developmental and Reproductive Toxicant (DART) Identification Committee. That DART Identification Committee evaluates selected agents for reproductive toxicity using a set of criteria adapted from the U.S. Environmental Protection Agency (EPA). The products of the evaluations are Hazard Identification Documents that can be found on the Internet at <http://www.oehha.org/prop65.html>. Thus far, only a few agents (methyl-tert-butyl ether, benzene, inorganic arsenic, and cadium) have been evaluated by the DART Identification Committee.

Reproductive and Developmental Toxicity

The DART Identification Committee evaluated each selected agent for male and female reproductive toxicity and developmental toxicity in humans and experimental animals. It also reviews other types of available toxicity data (e.g., acute and carcinogenicity) and pharmacokinetic data.

Quality Control

The documents are initially prepared by scientists with expertise in reproductive toxicity from the California EPA Reproductive and Cancer Hazard Assessment Section. They are then reviewed by the DART Identification Committee and open to public comment. The final documents are used to determine whether an agent should be placed on the state's list.

Usefulness in Identifying Exposures that Pose a Substantial Risk of Reproductive or Developmental Toxicity in Humans

The DART Identification Committee's evaluation is comprehensive in assessing reproductive toxicity. Although currently there are only a few documents available, they can be useful for providing a comprehensive review and evaluation for each agent.

Evaluative Process (Moore et al. 1995a,b; 1997)

Description

An expert committee has produced documents that assess lithium and boric acid for reproductive and developmental toxicity (J.A. Moore et al. 1995b, 1997). That committee used the evaluative process developed by Moore et al. (1995a); Chapters 2 and 3 of this report are based on that process. The reviews of lithium and boron follow a systematic approach for assessing agents for reproductive and developmental toxicity: (1) reviewing the conditions of exposure; (2) reviewing and summarizing the agent's toxicological and biological properties; (3) reviewing and summarizing human and experimental animal studies on reproductive and developmental toxicity; (4) integrating the reproductive and developmental toxicity data with information derived from the review of toxicological and biological properties, including pharmacokinetic data and exposure information; and (5) determining the certainty of the judgments about an agent's potential reproductive and developmental risk to humans. The expert committee's conclusions are stated in a summary.

Reproductive and Developmental Toxicity

The evaluative process was used to assess lithium and boric acid for their potential to cause reproductive and developmental toxicity (J.A. Moore et al. 1995b, 1997). The final product is a summary statement that integrates reproductive and developmental data and exposure data. The summary also makes a statement about the certainty of judgment about an agent's reproductive and developmental risk, and critical data needed for reducing uncertainty and improving the risk assessment.

Quality Control

The evaluative process was conducted by a team of scientists with expertise in a variety of disciplines related to reproductive and developmental toxicology.

Usefulness in Identifying Exposures That Pose a Substantial Risk of Reproductive and Developmental Toxicity in Humans

The evaluative process provides a comprehensive assessment of human reproductive and developmental toxicity of agent exposures.

National Institute of Environmental Health Sciences Center for the Evaluation of Risks to Human Reproduction

Description

Only recently established, the National Institute of Environmental Health Sciences (NIEHS) Center for the Evaluation of Risks to Human Reproduction will provide comprehensive, unbiased, and scientifically sound assessments of reproductive health hazards (including developmental health hazards) associated with human exposures to environmental agents. The assessment process to be used will be based on the evaluative process developed by J.A. Moore et al. (1995a). The center's product will be reports published in the peer-reviewed literature. The reports will be produced by expert panels in reproductive and developmental toxicology and reviewed by a committee composed of scientists from the center, NIEHS, and other institutions. Agents to be assessed are selected by an oversight committee composed of NIEHS scientists and representatives of federal and state agencies, public interest groups, and cofunders. Several phthalates have been reviewed and the assessments will be released to the public via the Center's website (http://cerhr.niehs.nih.gov). No other reports are available yet.

Reproductive and Developmental Toxicity

Reports produced by the center will assess environmental agent exposures for their potential to cause reproductive and developmental toxicity in humans. All relevant data are evaluated.

Quality Control

Center reports are prepared by an expert panel that comprehensively evaluates the relevant data, reviewed by a committee of scientists from NIEHS, the center, and other institutions, and approved by an oversight committee.

Usefulness in Identifying Exposures That Pose a Substantial Risk of Reproductive or Developmental Toxicity in Humans

The center reports will be comprehensive assessments of potential reproductive and developmental toxicity in humans for selected environmental agent exposures.

Informational Summaries

Reproductive Hazards of Industrial Chemicals: An Evaluation of Animal and Human Data

Description

Reproductive Hazards of Industrial Chemicals: An Evaluation of Animal and Human Data (Barlow and Sullivan 1982) is a classic reference book that provides detailed reviews of the reproductive and developmental toxicity of about 50 common occupational exposures. Each substance is treated in a separate chapter introduced by a description of the chemical, its general pharmacology, and its overall toxicity. Extensive citations to literature published before 1982 are included.

Reproductive and Developmental Toxicity

Reproductive Hazards of Industrial Chemicals provides detailed discussions of the full range of reproductive toxicity. Data from animal studies and investigations in humans are considered separately. Overall interpretations are provided, whenever possible, for reproductive endpoints in the context of general maternal toxicity.

Quality Control

S.M. Barlow and F.M. Sullivan are internationally-recognized authorities in reproductive and developmental toxicology. There is no formal external peer review of the book's contents.

Usefulness in Identifying Exposures That Pose a Substantial Risk of Reproductive or Developmental Toxicity in Humans

Reproductive Hazards of Industrial Chemicals is a valuable guide to information available prior to 1982 on the reproductive toxicity of the chemicals included. Unfortunately, the book is out of date and has not been revised.

REPROTEXT

Description

REPROTEXT is an electronic knowledge base developed by B. Dabney. It includes reviews of more than 850 commonly encountered industrial agents. Most are chemicals, but some physical agents are also covered. Subscriptions to REPROTEXT are available on CD-ROM from MICROMEDEX, Inc., as a component of its REPRORISK system.

Reproductive and Developmental Toxicity

REPROTEXT summaries provide information on the physical characteristics, use, and monitoring of agents as well as detailed discussions of the full range of general and reproductive toxicity. An extensive list of bibliographic citations is provided, reflecting the breadth of the material included. A grading system gives overall ratings of the general toxicity and reproductive hazard associated with each agent.

Quality Control

REPROTEXT summaries are written by an expert reproductive and developmental toxicologist. There is no formal external peer review of the summaries.

Usefulness in Identifying Exposures That Pose a Substantial Risk of Reproductive or Developmental Toxicity in Humans

REPROTEXT summaries provide a comprehensive overview of the information available on the reproductive and developmental toxicity of many important occupational exposures. The data are interpreted in the context of human reproductive hazards.

REPROTOX

Description

The REPROTOX knowledge base was developed by A. R. Scialli and associates at the Reproductive Toxicology Center, Bethesda, MD. REPROTOX is available on CD-ROM as part of the REPRORISK package distributed by MICROMEDEX, Inc. Subscriptions in other electronic formats are available through the Reproductive Toxicology Center. The knowledge base also has been published as a book, *Reproductive Effects of Chemical, Physical and Biologic Agents* (Scialli et al. 1995), but the electronic versions are more current.

Reproductive and Developmental Toxicity

REPROTOX has information on the reproductive and developmental toxicology of some 3000 chemical and physical agents. Each entry summarizes available data on a variety of reproductive outcomes in males and females. Each study is briefly described, and bibliographic citations are provided. The emphasis is on investigations in humans, but experimental animal data also are summarized. The primary intended audience is obstetricians and other physicians.

Quality Control

REPROTOX is written by clinical teratologists who have recognized expertise in the interpretation of reproductive and developmental toxicology studies in the context of actual human exposures. The summaries are revised, if necessary, in response to concerns raised by users and are frequently reviewed to ensure that the information included is up to date. There is no formal external peer review of REPROTOX summaries.

Usefulness in Identifying Exposures That Pose a Substantial Risk of Reproductive or Developmental Toxicity in Humans

REPROTOX is useful clinically. It provides succinct, authoritative overviews of the reproductive and developmental toxicity of a large number of chemical exposures, including many important occupational exposures. The system is designed for use by health professionals who are familiar with the general principles of risk assessment and is intended to assist in the counseling of individual patients.

Chemically Induced Birth Defects

Description

Chemically-Induced Birth Defects (Schardein 2000) describes mammalian teratology studies on more than 3300 chemicals, including those found in occupational exposures and in drugs. Each chapter deals with a group of related agents, such as industrial solvents. Literature citations are extensive.

Reproductive and Developmental Toxicity

Chemically-Induced Birth Defects contains information on human and other conventional mammalian teratology studies. Other kinds of reproductive toxicology are not usually covered. Full discussions are provided on chemicals for which substantial data exist, but most chemicals, for which information is much more limited, are listed in tables that summarize studies by species as "+" (developmental abnormalities observed), "-" (developmental abnormalities not observed), and "±" (equivocal results).

Quality Control

Chemically-Induced Birth Defects is written by an internationally-recognized authority in teratology. There is no formal external peer review of the contents.

Usefulness in Identifying Exposures That Pose a Substantial Risk of Reproductive or Developmental Toxicity in Humans

This book provides a useful overview of mammalian teratology studies on many important occupational exposures. When extensive data are available, they are interpreted in the context of human reproductive hazards.

Shepard's Catalog of Teratogenic Agents

Description

Currently in its eighth edition, H. Shepard's *Catalog of Teratogenic Agents* (1998) has been in print since 1973. Shepard's *Catalog* is also distributed in the MICROMEDEX, Inc. REPRORISK CD-ROM package and in other electronic formats through the TERIS Project at the University of Washington, Seattle. The Catalog is intended for use by investigators and clinicians.

Reproductive and Developmental Toxicity

Shepard's *Catalog* contains information on teratology studies of more than 2600 chemical and physical agents; other kinds of reproductive toxicity are not usually covered. Each entry summarizes available data from human and laboratory animal studies. Each study is described, and bibliographic citations are provided.

Quality Control

T.H. Shepard is an internationally recognized authority on clinical teratology. There is no formal external peer review of the summaries.

Usefulness in Identifying Exposures That Pose a Substantial Risk of Reproductive or Developmental Toxicity in Humans

Shepard's *Catalog* provides succinct, authoritative overviews of the information available on teratogenicity of many chemicals, including many that are important in occupational exposures. The data generally are not interpreted in the context of human reproductive hazards.

TERIS

Description

The Teratogen Information System (TERIS) is an electronic knowledge base developed by J. M. Friedman and J. Polifka. It includes information on more than 1000 agents, most of which are medications. TERIS is available on CD-ROM as part of the MICROMEDEX, Inc., REPRORISK package and in other electronic formats from the TERIS Project at the University of Washington, Seattle. The information also has been published as a book, *Teratogenic Effects of Drugs: A Resource for Clinicians (TERIS)* (Friedman and Polifka 2000); but the electronic versions are more current.

Reproductive and Developmental Toxicity

TERIS is designed to assist in counseling pregnant women about possible developmental risks; other forms of reproductive toxicity are not considered. In addition to brief critical reviews and references to the peer-reviewed literature, each TERIS summary contains an assessment of the risk of teratogenic effects in children of women with typical exposures during pregnancy. An evaluation of the quality of data available to determine the risk is also provided.

Quality Control

TERIS risk statements are based on consensus of the TERIS Advisory Board, a group of recognized authorities in clinical teratology. The board also provides peer review of all TERIS summaries.

Usefulness in Identifying Exposures That Pose a Substantial Risk of Reproductive or Developmental Toxicity in Humans

TERIS is useful clinically. It provides succinct, authoritative summaries and interpretations of information available on the teratogenicity of a large number of drugs and a few important occupational agents. No information is included on other forms of reproductive toxicity. The system is designed for use by health professionals who are familiar with the general principles of risk assessment and is intended to assist in the counseling of individual patients.

Bibliographic Resources

Developmental and Reproductive Toxicology Data Base

Description

The Developmental and Reproductive Database (DART) is a bibliographic resource maintained by the National Library of Medicine (NLM). DART has more extensive coverage of the reproductive and developmental toxicology literature than Medlars Online (MEDLINE), and its records are similar in format to those in MEDLINE. Each record provides bibliographic information, Medical Subject Headings (MeSH) indexing terms, Chemical Abstract Service (CAS) numbers, and an abstract (if available) for one publication in the open literature. DART contains more than 30,000 records for papers published since 1989; more than 3600 new records are added each year. DART captures approximately 60% of its records from MEDLINE; the remaining 40% represent journals not covered by MEDLINE, books, and meeting proceedings. DART is available on-line through the NLM's Toxicology and Data Network (TOXNET) system, on the Internet at <http://toxnet.nlm.nih.gov>.

Reproductive and Developmental Toxicity

DART provides bibliographic information and abstracts on reproductive and developmental toxicology studies. There is better coverage of developmental toxicity literature than reproductive toxicity
literature. The studies cover a broad range of species and reproductive
and developmental endpoints.

Quality Control

DART does not evaluate the studies included. Many of them, but
not all, have been subjected to conventional peer review prior to publication.

*Usefulness in Identifying Exposures That Pose a Substantial Risk of
Reproductive or Developmental Toxicity in Humans*

DART is an essential tool in identifying recent published literature
on reproductive and developmental toxicology, but considerable
expertise is required to interpret the information obtained in terms of
risks related to actual human exposures.

Environmental Teratology Information Center Backfile

Description

The Environmental Teratology Information Center Backfile
(ETICBACK) is a bibliographic resource primarily on developmental
toxicology studies published from 1950 to 1989 (before the DART
system was established). Each ETICBACK record provides bibliographic information, CAS registry numbers, and special keywords
related to species and treatment details. ETICBACK is available on-
line through NLM's TOXNET system: <http://toxnet.nlm.nih.gov>.

Reproductive and Developmental Toxicity

ETICBACK has more than 50,000 bibliographic records. The studies include a broad range of species and developmental toxicology endpoints.

Quality Control

ETICBACK does not evaluate the studies indexed. Many of them, but not all, have been subjected to conventional peer review prior to publication.

Usefulness in Identifying Exposures That Pose a Substantial Risk of Reproductive and Developmental Toxicity in Humans

ETICBACK is a useful tool in identifying older published literature on reproductive and developmental toxicology, but considerable expertise is required to interpret the information obtained in terms of risks related to actual human exposures.

Primary Data

National Toxicology Program Teratology Studies

Description

The purpose of the National Toxicology Program (NTP) Teratology Studies is to determine the developmental toxicity of chemicals to which women of child-bearing age, primarily pregnant women, are exposed. The studies are conducted in experimental animals and are designed to determine a dose-response relationship and to detect potential for developmental toxicity. This information can be used to provide a basis for human risk assessment. Requests to test chemicals come from the general public, from NIEHS personnel and other government agencies, international agencies, academia, and industry. Information, including a list of chemicals tested, can be found on the Internet at <http://www.niehs.nih.gov>. Individual documents can be ordered through the National Technical Information Service (NTIS).

Reproductive and Developmental Toxicity

These studies are designed to test chemicals for developmental toxicity. The program is also developing several techniques for evaluating potential toxic effects of chemical exposure on the reproductive system of humans and rodent models and on the developing embryos of rodents. Research efforts include increasing current knowledge about the site and mechanism of action for reproductive and developmental toxicity.

Quality Control

The results of all NTP studies undergo a rigorous public peer review. The study reports are evaluated by a standing subcommittee of the NTP Board of Scientific Counselors.

Usefulness in Identifying Exposures That Pose a Substantial Risk of Reproductive and Developmental Toxicity in Humans

The studies are designed to test selected exposures for developmental toxicity, but considerable expertise is required to extrapolate the findings to humans.

NTP Reproductive Assessment by Continuous Breeding Studies

Description

The purpose of the NTP Reproductive Assessment by Continuous Breeding (RACB) Studies is to identify potential hazards to male and/or female reproduction. The studies are conducted in experimental animals and are designed to determine a dose-response relationship and to detect potential for reproductive toxicity. This information can be used to provide a basis for human risk assessment. Chemicals may be nominated for testing through the Office of Chemical Nomination in the Environmental Toxicology Program at NIEHS or directly from the public to Reproductive Toxicology Group at NTP. Information, including a list of chemicals tested, can be found on the Internet at

<http://www.niehs.nih.gov>. Individual documents can be ordered through the National Technical Information Service (NTIS).

Reproductive and Developmental Toxicity

The RACB studies were designed by NTP to test chemicals for potential reproductive toxicity in males and females. In addition, this two-generation study design can be used to characterize the toxicity and to define the dose-response relationship for each chemical.

Quality Control

The results of all NTP studies undergo rigorous public peer review. The study reports are evaluated by a standing subcommittee of the NTP Board of Scientific Counselors.

Usefulness in Identifying Exposures That Pose a Substantial Risk of Reproductive and Developmental Toxicity in Humans

The studies are designed to test selected agent exposures for reproductive toxicity, but considerable expertise is required to extrapolate the findings to humans.

NTP Short-Term Developmental and Reproductive Toxicity Studies

Description

The purpose of the NTP Short-Term Developmental and Reproductive Toxicity Studies is to assess the general reproductive and developmental toxicities of substances identified by EPA as possible drinking-water contaminants. The general approach is designed to identify, using rats, the physiological processes (development, female reproduction, male reproduction, and effects on various somatic organs and processes) that are most sensitive to exposure to selected chemicals. Information, including a list of chemicals tested, can be found on the Internet at <http://www.niehs.nih.gov>. Individual documents can be ordered through NTIS.

Reproductive and Developmental Toxicity

The studies are designed using the NTP's Short-Term Reproductive and Developmental Toxicity Screen protocol (M.W. Harris et al. 1992). They are designed to provide preliminary data on the reproductive and developmental toxicity of chemicals about which little or no data exist. The results can be used to select chemicals for further study, to delineate the relative toxicities of structurally related chemicals, and to identify the proper dose range for subsequent toxicity studies.

Quality Control

The results of all NTP studies undergo rigorous public peer review. The study reports are evaluated by a standing subcommittee of the NTP Board of Scientific Counselors.

Usefulness in Identifying Exposures That Pose a Substantial Risk of Reproductive and Developmental Toxicity in Humans

The studies are designed to test selected agent exposures for reproductive and developmental toxicity, but considerable expertise is required to extrapolate the findings to humans.

ADDITIONAL SOURCES THAT MAY INCLUDE INFORMATION ON REPRODUCTIVE AND DEVELOPMENTAL TOXICITY

Detailed Evaluations

Agency for Toxic Substances and Disease Registry Toxicological Profiles

Description

The Agency for Toxic Substances and Disease Registry (ATSDR) toxicological profiles succinctly characterize the toxicological and adverse health effects information for hazardous substances. Each profile has a public health statement that describes toxicological properties in nontechnical terms. The profile also provides information about the amounts of significant human exposure and, where known, significant health effects for acute, subchronic, and chronic exposures. The adequacy of information also is described. Responsibility for the content and views expressed in a profile resides with ATSDR. A list of substances covered can be found on the Internet at <http://-www.atsdr.cdc..gov/toxpro2.html>.

The profiles are used by federal, state and local health officials and health professionals, interested private-sector organizations and groups; and members of the public. The agency's intent is to revise and republish profiles as newer data become available.

Reproductive and Developmental Toxicity

Pertinent information on reproductive and developmental effects is presented in a narrative summary and in tabular form for species, route, dose, no-observed-adverse-effect level (NOAEL), and lowest-observed-adverse-effect level (LOAEL).

Quality Control

Draft profiles are usually prepared by contractors with demonstrated technical expertise. The drafts are reviewed by several groups within ATSDR, published for public comment, and peer reviewed by several experts. ATSDR evaluates peer reviewers' comments, and comments not incorporated in the profile are made part of the administrative record with a brief explanation of the rationale for their exclusion.

Usefulness in Identifying Exposures That Pose a Substantial Risk of Reproductive or Developmental Toxicity in Humans

Information presented in the profiles can be useful for identifying and summarizing studies relevant to reproductive and developmental toxicity. Significant exposures to substances are reported.

National Institute for Occupational Safety and Health Criteria Documents

Description

The National Institute for Occupational Safety and Health (NIOSH) develops criteria documents to describe the scientific basis for occupational safety and health standards. They contain critical reviews of the available literature on physical and chemical properties, uses and occurrence, toxicokinetics, general toxicity, toxic effects on various organs, genotoxicity, carcinogenicity, and developmental and reproductive toxicity of particular agents. Data are evaluated in the context of potential human occupational exposures, and recommendations for minimizing safety and health risks are provided. Most of these documents were written more than 10 years ago, and many are more than 20 years old.

Criteria documents are developed primarily for the U.S. Department of Labor, but they are also distributed to health professionals, industry, organized labor, public interest groups, and federal, state, and local government agencies. More than 140 NIOSH criteria documents are available on the Internet at <www.cdc.gov/niosh>. Individual documents or the entire set on CD-ROM can be obtained from NTIS.

Reproductive and Developmental Toxicity

Available information on reproductive and developmental effects is presented and critically evaluated. Human data and animal studies are considered separately and assessed in terms of potential implications for human occupational exposures.

Quality Control

NIOSH criteria documents are developed by NIOSH staff and reviewed by external expert consultants.

Usefulness in Identifying Exposures That Pose a Substantial Risk of Reproductive or Developmental Toxicity in Humans

NIOSH criteria documents provide information that is useful for identifying exposures that might be associated with reproductive and developmental toxicity. Unfortunately, most of the documents are out of date.

International Agency for Research on Cancer Monographs on the Evaluation of Carcinogenic Risks to Humans

Description

These monographs are prepared and published by the International Agency for Research on Cancer (IARC), which is part of WHO. Qualitative evaluations of published literature are made by international working groups of independent scientists. A list of agents covered can be found on the Internet at <http://www.iarc.fr>.

Reproductive and Developmental Toxicity

Each agent evaluation has a section that briefly summarizes reproductive and developmental effects using the traditional IARC format of presenting human data separately from data in experimental systems.

Quality Control

IARC working groups are composed of individuals with expertise related to cancer. There usually is a member with expertise on reproductive and developmental toxicity who prepares that section, but the working group is not constituted to provide a peer review of reproductive or developmental toxicity.

Usefulness in Identifying Exposures That Pose a Substantial Risk of Reproductive or Developmental Toxicity in Humans

Information presented in the IARC evaluations can be useful for identifying and summarizing studies relevant to reproductive and developmental toxicity.

International Programme on Chemical Safety Environmental Health Criteria Documents

Description

The International Program on Chemical Safety (IPCS) has been preparing and issuing Environmental Health Criteria Documents since 1976. More than 170 documents have been issued. IPCS is a joint effort of the United Nations Environmental Program (UNEP), International Labor Organization (ILO), and WHO. The documents provide critical reviews on the effects on human health and the environment of chemicals, or combinations of chemicals, and physical and biological agents. The documents assist national and international authorities in making risk assessments and risk management decisions. Information about the documents, including a list of agents covered, can be found on the Internet at <http://www.who.int/pcs/>.

IPCS Environmental Health Criteria Documents reflect the collective view of an international group of experts and do not necessarily represent the decision or stated policy of UNEP, ILO, or WHO. The reports summarize and interpret the pertinent published literature. Unpublished information is used when published information is absent or when data are pivotal to the risk assessment. Adequate human data are preferred to animal data. Topics include physical chemistry; sources of exposure; environmental fate and transport; concentrations in the environment and in humans; pharmacokinetics; effects from acute, subacute, and long-term exposure; toxic effects on skin, eye, and reproduction; mutagenesis; and cancer.

Reproductive and Developmental Toxicity

Summary statements are provided under the heading of reproductive toxicity, embryotoxicity, and teratogenicity. Data from individual studies usually are not presented.

Quality Control

Typically, a draft is prepared by an expert and presented to an IPCS official. The draft is communicated to approximately 150 contact persons who have an opportunity to provide comments. A revised draft is then prepared and submitted to a technical group of international experts. Further changes are made to reflect their deliberations, and, after editing and approval by the director of IPCS, the document is issued.

Usefulness in Identifying Exposures That Pose a Substantial Risk of Reproductive or Developmental Toxicity in Humans

The Environmental Health Criteria Documents provide information on reproductive and developmental effects for selected agents that can be useful for identifying important studies.

EPA Integrated Risk Information System

Description

The Integrated Risk Information System (IRIS), prepared and maintained by EPA, is an electronic knowledge base containing information on human health effects that might result from exposure to various chemicals in the environment. IRIS was initially developed for EPA staff in response to a growing demand for consistent information on chemical substances for use in risk assessments, decision-making, and regulatory activities. The information in IRIS is intended for those without extensive training in toxicology but with some knowledge of health sciences. IRIS includes chronic reference doses (RfDs), reference concentrations (RfCs), and cancer unit risks and slope factors for more than 500 substances. It is searchable on the Internet at <http://-www.epa.gov/iris>.

Reproductive and Developmental Toxicity

The review of data for setting chronic RfDs and RfCs is comprehensive and includes a review of reproductive and developmental toxicity data. Two prenatal developmental toxicity studies in two species and a two-generation reproduction study are considered part of the minimum database for setting the chronic RfD-RfC. If such studies are not available, a database deficiency factor (ranging from 1 to 10) may be applied.

Quality Control

Draft IRIS summaries and support documents undergo rigorous internal and external peer review, followed by an agency-scientific consensus review. The IRIS program manager oversees the consensus process and provides quality control of all documents. A senior science manager provides scientific authority.

Usefulness in Identifying Exposures That Pose a Substantial Risk of Reproductive or Developmental Toxicity in Humans

Information in the files for the RfD and RfC can be useful for identifying a risk of reproductive and developmental toxicity in humans, if such data have been used in setting the RfD or RfC. The review of data in establishing RfDs and RfCs is comprehensive, and newer assessments have detailed support documents that can be downloaded from the IRIS Internet site, <http://www.epa.gov/iris>.

EPA National Center for Environmental Assessment Documents

Description

The National Center for Environmental Assessment (NCEA), which is part of the EPA Office of Research and Development (ORD), serves as the national resource center for the overall process of human health and ecological risk assessments; the integration of hazard, dose-response, and exposure data; and models to produce risk characterizations. NCEA prepares a variety of documents, many of which are the source of scientific information used by EPA decision makers in developing or revising regulations. The documents can pertain to a specific medium, such as air or water, or they can be comprehensive analyses of scientific data. Many NCEA documents contain analyses of reproductive and developmental toxicity data. The documents are available on the NCEA web site, http://www.epa.gov/ncea, and from NCEA's technical information staff.

Reproductive and Developmental Toxicity

Available reproductive and developmental toxicity data are reviewed and summarized.

Quality Control

All NCEA documents undergo internal and external peer review according to EPA, ORD, and NCEA policies. Each document is cleared by a senior manager before publication.

Usefulness in Identifying Exposures that Pose a Substantial Risk of Reproductive and Developmental Toxicity in Humans

The review of data is comprehensive, and information in the documents can be useful for identifying a risk of reproductive and developmental toxicity in humans.

Informational Summary

Chemical Hazards of the Workplace

Description

Proctor and Hughes' Chemical Hazards of the Workplace (Hathaway et al. 1996) is a standard reference that provides summaries of the toxicity of more than 500 frequently encountered occupational chemicals. Each summary includes American Conference of Governmental Industrial Hygienists Threshold Limit Values (when available); a statement describing physical properties, uses, and usual routes of exposure; and selected literature citations. Brief descriptions of the clinical effects of toxic exposure are given along with the doses at which the effects have been observed. Emphasis is placed on data obtained from human studies.

Reproductive and Developmental Toxicity

The summaries for chemical exposures that have been associated with human reproductive or developmental toxicity usually include a paragraph describing those effects.

Quality Control

Chemical Hazards of the Workplace is written by internationally recognized authorities in occupational toxicology. There is no formal external peer review of the summaries.

Usefulness in Identifying Exposures That Pose a Substantial Risk of Reproductive or Developmental Toxicity in Humans

The book provides a short, authoritative overview of the information available on the occupational toxicology of many important chemical exposures. Consideration of reproductive and developmental toxicology is limited.

Bibliographic Resources

Bibliographic resources that may contain information on reproductive and developmental toxicity include the NIOSH Registry of Toxic Effects of Chemical Substances (RTECS), TOXLINE, MEDLINE, and PUBMED. RTECS is a NIOSH database of toxicological information—including mutagenic effects, reproductive effects, tumorgenic effects, and other toxicity—obtained from the open scientific literature. RTECS includes study summaries (route, species, study type, dose, effect) and bibliographic information on reproductive effects, if available. RTECS does not evaluate the studies included. Considerable expertise is required to interpret the information obtained in terms of risks related to human exposure. Originally published in 1971 as the "Toxic Substances List," RTECS contains information on more than 130,000 substances. It is available on the Internet at http://www.cdc.gov/niosh/rtecs.html, on CD-ROM, and on computer tape.

TOXLINE is NLM's collection of online bibliographic information covering the biochemical, pharmacological, physiological, and toxicological effects (including effects on reproduction and development) of drugs and other chemicals. It contains more than 2.5 million bibliographic citations, most with abstracts and/or indexing terms and CAS Registry Numbers. It is available on the Internet at http://-toxnet.nlm.nih.gov/. MEDLINE is an NLM bibliographic database covering the fields of medicine, nursing, dentistry, veterinary medicine, the health care system, and the preclinical sciences. MEDLINE contains bibliographic citations and author abstracts from more than 4,000 biomedical journals published in the United States and 70 other countries. The file contains over 11 million citations dating back to the mid-1960's. MEDLINE can be accessed on the Internet at: http:// igm.nlm.nih.gov/. PubMed was developed by the National Center for Biotechnology Information (NCBI) at NLM. PubMed provides access to bibliographic biomedical-related information, which is drawn primarily from MEDLINE, PreMEDLINE, HealthSTAR, and publisher-supplied citations. PubMed also provides access and links to the integrated molecular biology databases included in NCBI's Entrez retrieval system. These databases contain DNA and protein sequences, 3-D protein structure data, population study data sets, and assemblies of complete genomes in an integrated system. PUBMED can be accessed on the Internet at: http://www.ncbi.nlm.nih.gov/entrez/query.fcgi.

Primary Data

National Toxicology Program Toxicology Report Series

Description

Acute or repeated dose studies of up to 90 days are conducted in rats and mice on selected agents under the direction of NIEHS. Results from genetic toxicity testing and pharmacokinetic studies are frequently presented.

Reproductive and Developmental Toxicity

Reports of 90-day repeated-dose studies usually incorporate an evaluation of sperm morphology and vaginal cytology. The procedures are performed on 10 animals per dose group. Data from males are collected at time of necropsy. For females, vaginal swabs are collected daily for the last 12 days of the study. Typically, tables for males include data for weight of the left testis, left epididymis, and cauda epididymis and data on sperm count, concentration, and motility. Data from females also are expressed as mean estrus cycle length and the percent of cycle in diestrus, proestrus, estrus, or metestrus. It is standard for a histological examination of testis and ovary to be performed in each study on animals from the highest dose groups and from control groups. Lower dose groups may be evaluated when effects are observed at the high dose.

Quality Control

Studies since the mid-1980s have generally been performed in conformance with good laboratory practice regulations. Histological examinations are reviewed by a pathology working group. Each draft report is reviewed independently. The reviewers determine whether the design and conditions of the studies are appropriate and ensure that toxicity reports present results and conclusions fully and clearly. The comments of the reviewers are not summarized or otherwise presented in reports, but the comments of peer reviewers are reviewed and addressed.

Usefulness in Identifying Exposures That Pose a Substantial Risk of Reproductive or Developmental Toxicity in Humans

The NTP toxicology reports can be useful for identifying agents that cause male or female reproductive (but not developmental) effects in those parameters examined.

Organization for Economic Cooperation and Development Screening Information Data Set Profiles

Description

The Organization for Economic Cooperation and Development (OECD) has committed to the development and assessment of a core set of toxicological and environmental data on 2000 chemical substances with high production and use throughout the world. The Screening Information Data Set (SIDS) represents the core data that are assessed to determine whether a substance can be set aside with low priority for additional evaluation or needs further study because of possible environmental or human health effects. The data address physical and chemical properties and acute and repeated dose (28-day) mammalian toxicity. SIDS dossiers are reviewed by a committee of technical and scientific experts from OECD-member countries. More than 100 dossiers have been completed. Information on SIDS can be found on the Internet at <http://www.oecd.org/ehs/hpv.htm>.

Reproductive and Developmental Toxicity

The core data requirements include assessment of fertility and reproduction. This requirement is commonly met through a 28-day repeated-dose protocol in rats, which requires treated members of both sexes to be mated and females observed for pregnancy, parturition, and survival of pups through postnatal day 4. Histological evaluations of testis and ovary also are required. Positive test results are not considered definitive, in part because of the limited numbers of animals used, but they signal the need to confirm and expand the characterization of possible effects using more traditional protocols.

Quality Control

The studies are done according to a standardized OECD protocol and in compliance with good laboratory practices. Test results are reviewed by the responsible government entity within the country with lead responsibility. Dossiers are subsequently reviewed by an international committee of experts.

Usefulness in Identifying Exposures That Pose a Substantial Risk of Reproductive or Developmental Toxicity in Humans

The SIDS profiles rank chemical substances based on their potential environmental and human health effects. Some information on reproductive and developmental toxicity is included in each profile.

Appendix C

Human Study Designs

Broadly, epidemiological studies can be categorized into three types: descriptive studies that focus on the occurrence of disease or health-related states in specific populations or representative samples, analytical studies designed to assess associations or test hypotheses about risk factors or exposures and health outcomes, and experimental trials in which investigators randomly assign exposures to treatment groups. Table C-1 lists the available designs by type of study.

TABLE C-1 Epidemiological Research Design

Descriptive	Analytical	Experimental
Case series	Case-control	Community
Cross-sectional	Cohort	Randomized Clinical
Ecological	• Retrospective	
	• Prospective	

Descriptive studies do not formally test hypotheses; rather, they generate hypotheses based on evaluation of research questions. As such, descriptive designs cannot assess causality. One common descriptive design that offers considerable information on selected outcomes, such as birth defects, is the case series design. As the name suggests, this design encompasses a series of cases with the same outcome. There is no comparison group. This type of study can raise suspicion of an association and, in fact, has been instrumental in identifying certain adverse effects (e.g., diethylstilbestrol and vaginal adenocarcinoma). In fact, much of the available data on adverse outcomes and pharmaceutical compounds comes from case series designs. Cross-sectional studies measure exposures and outcomes at the same point in time. Correlational or ecological studies attempt to correlate an exposure with an outcome. Individual case studies or case series also are used. Most descriptive studies compare disease or health-related endpoints in relationship to a specific exposure or risk factor. Because comparison groups vary with regard to other factors associated with the exposure, further assessment of associations is needed, and causality cannot be determined.

Analytical studies include cohort (prospective and retrospective) and case-control (retrospective) types. Several hybrid designs exist as well, such as those that use retrospective cohorts. Control studies might be matched or unmatched in the design phase; matched-cohort studies are relatively rare, despite offering improved efficiency over other designs (K.J. Rothman and Greenland 1998). The major distinction between cohort and case-control designs is that cohort studies begin with the exposure and follow individuals to ascertain incident or new cases of disease. In this regard, the investigator has confidence in the temporal ordering between exposure and outcome. Case-control studies, on the other hand, start with disease status and retrospectively ascertain exposure, and may be subject to biases associated with the correction of data.

Experimental designs include randomized clinical (or community) trials and are considered the most scientifically desirable design available to epidemiologists. These designs ensure the temporal ordering between an exposure and outcome and minimize confounding via the randomization process by maximizing the internal validity of findings; external validity may be limited. Such designs have limited applicability to environmental and occupational epidemiology, given that exposures typically cannot be randomly assigned. Few "natural" experiments occur in which a particular subgroup of the population is exposed while others are not, and exposure is not randomized in such instances.

Scientifically sound epidemiological studies adhere to the essential elements of the epidemiological method:

1. Formulation of a well-defined research question or study hypothesis suitable for testing.
2. Description of the referent population or representative (probability) sample;
3. use of standardized methodology for data collection (exposure, outcome, effect modifiers, confounders).
4. Application of a well-described and appropriate analytical plan.
5. careful interpretation of the data using an established paradigm for assessing causality.

After carefully weighing the choice of study design, the existing literature should be used to ground the hypothesis within a theoretical framework to enhance biological plausibility in interpreting the results. Characterization and selection of the study population or representative (probability) sample is extremely important and requires careful consideration. Random-sampling techniques should be used to ensure that each individual in the referent population has an equal chance of being selected. That approach minimizes bias and thereby enhances the validity of findings. Careful attention must be given to the inclusion and exclusion criteria that can render a sample too restrictive, resulting in limited external validity (generalizability). The effect of occupational exposures on reproductive and developmental outcomes is of added interest in that employed females might include a higher proportion of sub- or infertile women than is found in the general population. If fertile women leave the work force for child-bearing, bias can be introduced into the study, resulting in the "infertile worker effect" (Joffe 1985).

Use of a standardized methodology for data collection is critical for collecting information for all study subjects, regardless of exposure or disease status. Information must be collected on exposure, outcome, and effect modifiers or confounders. There are several methods available for ascertaining information on exposures and outcomes, such as self-reported data obtained in personal or telephone interviews, self-administered questionnaires, diaries, observation, existing records, actual physical measurements, and collection of biological specimens (Armstrong et al. 1995).

National and state registries provide another source of data that can be used for epidemiological studies that assess the effects of particular exposures. All states maintain records in fetal death and live birth and death registries, and all are population based. Some states have cancer or birth defects registries. With the exception of live birth and death registries, states vary in their mechanisms (active versus passive) and requirements for surveillance. There also is variability in which developmental defects are ascertained and how they are classified. Hence, registry data are not all comparable. Most United States registries do not have exposure data readily available for analysis of outcomes, so this information must be collected retrospectively or surrogate information, such as parental occupation or residence location, must be used. For European countries with centralized health care systems, some prospectively collected exposure data can be linked to other registries, such as birth defect or live birth registries. The increasing frequency of pregnancy termination when prenatal diagnosis detects fetal anomalies could underestimate the accuracy of registry data. Hence, the birth prevalence of malformations may be estimated, but the incidence remains unknown.

Pregnancy registries established for postmarketing surveillance of pharmaceutical substances could offer some information for assessing reproductive and developmental toxicity. This data source could have limited validity, however, because only a proportion of affected or exposed women are included in such registries. The representativeness of registry data will be determined in part by the prevalence of the exposure across the population at risk, the voluntary nature of the registry, the type of sponsor (industry, government, university), and reporting and surveillance mechanisms. Similar concerns could affect data obtained by following-up on women who contact teratogen information services because of concerns about possible exposures.

Registries could offer some preliminary information about the distribution and determinants of a few reproductive and developmental outcomes (fetal death, live births, birth defects), but often additional information on exposures and the precise nature of the adverse outcomes will need to be collected. Registry data are simply not available for most fecundity-related outcomes indicative of male and female reproductive health (conception delay, early pregnancy loss).

Exposure data varies in quality across epidemiological studies, especially those concerned with environmental exposures. Many earlier epidemiological studies relied exclusively on self-reported or proxy exposures (e.g., residence). For example, a study of the effect of air pollution on respiratory health might have compared ambient-air concentrations for pollutants by rate of respiratory disease. Without individual measurements, it is hard to know who was truly exposed (or not exposed). Of late, there is a growing trend to collect biological specimens suitable for estimating exposure. If there are inadequate resources for measuring exposures for all study participants, epidemiologists often will stratify subjects by estimated exposure and randomly select subjects for more detailed study of exposure status (with biological specimens). Simulation techniques can be used to evaluate how well the associations from exposure biomarker studies are upheld as theoretical sample sizes are increased.

Other important concerns with respect to exposure assessment include ensuring the temporal ordering of the exposure-to-outcome relationship and assessment of dose-response effects. If effects are interpreted as causal, then the temporal ordering of exposure to outcome must be established. A spectrum of reproductive and developmental outcomes might be possible, depending on critical windows of development. For example, a fetus exposed to thalidomide in the first trimester is at increased risk for phocomelia; the same exposure in the last trimester does not increase that risk. Although evidence of a dose-response relationship is important for assessing causality, often such relationships are lacking. Moreover, an inverse dose-response relationship might result in a high early pregnancy loss rate (or sterility) for those most heavily exposed and, thereby, appear to have a protective effect on risk of malformations or other adverse pregnancy outcomes. Consideration of potential fertility bias (Weinberg et al. 1994) is needed. Continuous quality control ensures the validity and reliability of data, and often a proportion of individuals are selected for a formal study of validity (e.g., confirmation with another data source, such as medical records or biomarkers) and reliability (i.e., individuals are queried at different points in time about exposures).

The fourth element of the epidemiological method includes a well-developed analytical plan that might be modified over the course of study. The plan must be appropriate for the study design and hypothesis under study, type of data collected and scale of measurement, completeness of data (e.g., percentage of missing data), distributions of variables, appropriateness of assumptions that underlie statistical techniques for the data set, consideration of potential effect modification or confounding, and statistical significance testing for sample data.

If multiple comparisons are made, it might be necessary to adjust for them (e.g., by Bonfierroni procedures) to ensure the validity of the findings.

The last step in the epidemiological method is the careful interpretation of findings. All alternative explanations (chance, random error, bias) must be considered carefully and eliminated in assessing causality. Negative findings should receive the same careful consideration as positive findings. To that end, a priori power estimates are extremely useful for determining the statistical power of the study and for assessing Type I and II errors.

Interpretation of findings requires evaluation of bias (systematic distortion), random error (noise), confounding (distortion produced by a third factor associated with both exposure and outcome), synergism (interaction of two or more causal factors to produce effects greater than the sum of individual effects) and effect modification (direction and strength of an association depending on a third variable) (Jekel et al. 1996). Bias is a major threat to validity that weaken or distort a true relationship between an exposure and disease or even produce a spurious one. Common sources of bias include the selection of study participants (selection bias), sources of information (information bias), or misclassification of subjects either by disease or by exposure status (misclassification bias). If subjects are randomly misclassified, effects will be underestimated (biased toward null). That might result in erroneous and negative findings. However, nonrandom or differential misclassification bias might produce effects that are either over- or underestimated. There are few statistical techniques for addressing bias (e.g., covariance adjustment and causal modeling); however, minimizing bias in the design phase is preferable to posthoc statistical adjustments.

Random error can over- or underestimate risk and is generally not as severe as bias. Moreover, the magnitude of error can be estimated with statistical techniques. Assessment of confounding, synergism, or effect modification can be accomplished in the analytical phase (by stratification or multivariate modeling), providing sufficient data have been collected on those factors. Restriction or randomization procedures also can be used in the design phase to minimize confounders.

Causality can be considered in analytical or experimental epidemiological studies. That involves assessing the statistical association, the temporal relationship between exposure and outcome, and the elimination of other potential explanations such as chance or bias (Jekel et al. 1996). There are several different criteria for assessing causality; Hill's criteria are cited often (Hill 1965). The existing paradigms for assessing causality have been discussed by Weed (1995) and are illustrated in Table C-2. Moreover, authors vary in degree to which they use criteria for assessing causality (Weed 1997). Despite explicit criteria for evaluating causality, scientists might vary in their use or interpretation of the criteria. Scientists must consider formalized strategies for weighing scientific evidence to assist in the interpretation of available information (Weed 1997).

TABLE C-2 Epidemiological Considerations Important for Causal Inference (1959-1973)

Lilienfeld (1959)	Sartwell (1960)	Surgeon General (1964); Susser (1973)	Hill (1965)	MacMahon and Pugh (1970)
Consistency	Replication	Consistency	Consistency	Strength of association (including magnitude of association and dose response)
Magnitude of effect	Strength of association	Strength of association (including magnitude of effect of dose-response)	Strength of association	
Dose-response	Dose-response		Biological gradiant	Temporality
Experimentation	Temporality	Temporality	Temporality	Experimentation
Biological mechanism	Biological reasonableness	Biological coherence	Experimental evidence	Consonance with existing knowledge
		Specificity	Biological plausibility	Biological mechanism
			Biological coherence	Consistency
			Specificity	Exclusion of alternative explanations
			Analogy	

Source: Adapted from Weed 1995.

Appendix D

Experimental Animal and In Vitro Study Designs

Experimental animal studies should be evaluated as part of hazard characterization to ensure that adequate research has been carried out. The design (choice of species, vehicle, route and timing of exposure), conduct, interpretation, and reporting should be considered. In any assessment of the reproductive and developmental toxicity potential of exposure to a potentially harmful substance, all available data should be considered, including supplementary data from studies that are not designed to test reproductive and developmental toxicity. Supplementary information can be obtained from acute (single or multiple exposures that occur within 24 hours or less), subchronic (multiple or continuous exposures that last up to 3 months), and chronic (multiple exposures that occur over a significant fraction of an animal's life span) systemic toxicity studies (particularly where reproductive organs have been examined) and from toxicokinetic or tissue distribution data. In vitro test systems also can provide information about an agent's potential to cause reproductive or developmental toxicity. By themselves, however, in vitro tests are insufficient for defining the potential reproductive or developmental toxicity of an agent.

The primary information on experimental animal testing for reproductive and developmental toxicity potential is likely to be derived from standard studies used by regulatory agencies. Several statutes and guidelines have been published by different authorities, such as the Organization for Economic Cooperation and Development (OECD 1983, 1984, 1995, 1996, 2000a,b), the U.S. Environmental Protection Agency (EPA 1998a,b,c,d), and the U.S. Food and Drug Administration (FDA 1994,2000).

This appendix describes experimental animal and in vitro studies that are used to assess developmental toxicity and male and female reproductive toxicity from exposures to pesticides, industrial chemicals, and food ingredients. The testing of pharmaceutical agents is not described in detail here, but can be found in FDA (1994). A summary of the study types, protocols, endpoints and limitations is presented in Table D-1. A description of the manifestations of each type of toxicity and guidance on the interpretation of results from the studies also are presented.

DEVELOPMENTAL TOXICITY

Developmental toxicity is defined as adverse effects in the developing organism that can result from exposure before conception in either parent, exposure during gestation, or exposure during postnatal development from birth to sexual maturation. Adverse developmental effects can be detected at any point in the life span of the organism. The major manifestations of developmental toxicity include death of the developing organism, structural abnormality, altered growth, and functional deficiency (EPA 1991).

Structural abnormalities in development include malformations and variations. A malformation is usually defined as a permanent structural change that can adversely affect survival, development, or function. The term *variation* indicates a divergence from the usual range of structural constitution that might not adversely affect survival or health. Because there is a continuum of responses from normal to severely abnormal, distinguishing between variations and malformations can be difficult.

Altered growth can result in an alteration in the size or weight of an organ or in body weight or size of exposed offspring. Changes in one indicator of altered growth might or might not be accompanied by other signs of altered growth. For example, changes in body weight

TABLE D-1 Types of Animal Studies Used to Assess Reproductive and Developmental Toxicity

Study Type	Purpose	Protocol	Endpoints	Limitations	Reference
Single-generation reproduction	Provides basic information on potential of agent to produce adverse effects on male and female reproduction	Before mating, males exposed for complete cycle of spermatogenesis and epididymal transit time; females exposed for at least 2 estrous cycles	Toxic effects, mortality, neurobehavioral changes, altered sexual behavior, problems in parturition and lactation, time to positive sperm smear, duration of pregnancy, number and sex of pups, stillbirths, live births, presence of gross abnormalities, weight changes, and structural abnormalities	Does not provide information on breeding capacity of the F_1 generation, effects expressed after weaning, individual male and female effects, reproductive senescence, reversibility, other specific functional developmental effects, time of effect initiation, structural anomalies of offspring (generally), internal dose	OECD 1983
Multi-generation reproduction	Determines potential of agent to produce adverse effects on male and female reproductive systems and on the	Before mating, males and females exposed 10 wk; offspring exposed through lactation and after weaning, by individual treatment;	Libido; estrous cyclicity, ovarian histopathology including quantification of primordial follicles in P and F_1 females; sperm parameters (count, motility, morphology) in P and F_1 males; fertilization; implantation; embryonic, fetal, neonatal growth and development; parturition; lactation; post-	Does not provide information on reproductive senescence, reversibility, detailed functional developmental effects other than on the reproductive system, time of effect initiation, structural anomalies of offspring (generally)	OECD 2000b; FDA 2000; EPA 1998b

| | | embryo, fetus, and neonate | dosing of all generations is continuous throughout study | weaning growth and sexual maturity; development of reproductive organs; brain, spleen, and thymus weights | | |
| Prenatal developmental toxicity | Determines potential of agent to produce adverse effects on animals exposed during gestation | Animals (rodent and nonrodent) exposed at least from implantation until just prior to parturition | | External visceral and skeletal malformations and variations, weight, pre-implantation and post-implantation loss | Does not provide information on reversibility and repair of specific effects, malformed offspring may die before observation, low power for detecting malformations, treatment does not usually cover period before implantation, function of fetal organs not evaluated, restricted macroscopic examination, specific susceptible period of development can not be identified, limited evaluation of maternal and adult toxicity | OECD 2000a; EPA 1998a; FDA 2000 |

TABLE D-1 (Continued)

Study Type	Purpose	Protocol	Endpoints	Limitations	Reference
Developmental neurotoxicity	Assesses potential neurotoxicity from exposure during critical stages of development	Pregnant animals exposed during gestation and lactation to postnatal day 10	Offspring tested for gross neurological and behavioral disorders, motor activity, response to auditory startle, learning and memory, neuropathological effects, brain weight	Exposure over whole period of postnatal development is not included, limited assessment of learning and memory	EPA 1998c
Serial mating (Dominant lethal)	Assesses stages of spermatogenesis and cell types in male reproduction for sensitivity to an agent	Adult males exposed for 1-5 d before mating then mated to 1-3 females weekly for 8-10 wk	Number of implantation sites in uteri, early fetal mortality	Does not provide information on reversibility, several general reproduction parameters as stated under the limitations section of the single- and two-generation reproduction studies, and endpoints of male reproductive toxicity	OECD 1984; EPA 1998d
Continuous breeding (RACB)	Similar to multigeneration except that reproductive capacity over a 14-wk	Males and females (rats and mice) treated for 1 wk before mating and for 14-wk mating	Same as those in multigeneration studies, as well as time between litters and progressive effects on fertility and reproduction	Does not provide information on individual male and female parental effects; reversibility; specific functional developmental effects in offspring, time of effect initiation, internal and	Reviewed in Lamb 1985

		period is also assessed	period. Litters removed at birth, examined, and discarded, except for last litter which is weaned, raised to breeding age, and mated to evaluate effects in second generation	skeletal anomalies in offspring, sexual behavior	
Total reproductive capacity	Assesses ovarian toxicity	Females (usually mice) exposed to an agent for a short period in utero or postnatally and allowed to mate with a single male while female remains fertile	Number of litters and offspring	Does not provide information on reversibility types of toxicity other than ovarian, effects in males	McLachlan et al. 1981; Generoso et al. 1971

sometimes accompany changes in crown-rump length or skeletal ossification. Altered growth can occur at any stage of development, and it can be reversible in some cases or permanent in others. Most current study designs do not allow differentiation between reversible and permanent changes.

Functional developmental toxicity is the study of alterations or delays in the physiological or biochemical competence of an organism or organ system after exposure to an agent during pre- or postnatal development. In any given test animal, delayed development can be assessed in relation to established landmarks for physical, behavioral, and sexual maturation.

Types of Studies

Two types of studies specifically designed to assess developmental toxicity are discussed in this section: the prenatal developmental toxicity study and the developmental neurotoxicity study. Several other types of studies, although not solely designed to assess developmental toxicity, can be used for that purpose. They include single- and multigeneration reproduction studies, reproductive assessment by continuous-breeding studies, and serial mating (dominant lethal) studies discussed in later sections.

Prenatal Developmental Toxicity

The prenatal developmental toxicity study provides information on the effects of repeated exposure to an agent during pregnancy (OECD 2000a; EPA 1998a; FDA 2000). It is normally conducted in two species, a rodent (usually rat) and a nonrodent (usually rabbit), although not all guidelines specify nonrodents. Animals are exposed to an agent, usually via ingestion or inhalation, during the period of major organogenesis. The protocols include exposure to the end of gestation in order to cover developmental events that occur later in gestation (e.g., central nervous system, skeletal growth, sexual differentiation). Offspring are delivered by cesarean section on the day before the expected day of parturition, and a maternal necropsy is conducted, including examination of the uterus for number of implantations, resorptions, fetal deaths, and live fetuses. Corpora lutea in the ovaries are also counted. Live fetuses are weighed and examined carefully for external, visceral, and skeletal malformations and variations. Although the terminology used for malformations and variations has been variable from laboratory to laboratory, attempts have been made at standardization (Wise et al. 1997).

Developmental Neurotoxicity

The objective of developmental neurotoxicity studies is to assess the potential of an agent to affect neurodevelopment (EPA 1998c). The protocol is designed to be used either as a separate study, usually as a follow up to other studies, or as part of a multigeneration reproduction study. A test agent is administered at a minimum of three dose levels to pregnant animals in groups that are large enough to produce 20 litters per dose group from day 6 of gestation through day 10 postnatally (the first half of lactation). (This is the minimum exposure period. Dosing can be continued throughout lactation or, in the context of a multigeneration study, dosing is done daily over two generations.) Pregnant and lactating dams are assessed for clinical signs of neurodevelopmental effects and for their performance in a functional observation battery. Litter sizes can be adjusted by random selection to provide equal numbers of male and female offspring (usually four of each). Offspring are randomly selected from litters for neurotoxicity evaluation, including gross neurological and behavioral disorders, motor activity, response to auditory startle, learning and memory, brain weight, and neuropathological examination. Motor activity is studied on postnatal days 13, 17, 21, and 60 ± 2. Auditory startle tests are conducted on postnatal days 22 and 60 ± 2. Learning and memory are evaluated in the offspring around the time of weaning (postnatal day 21) and again in adulthood (postnatal day 60 ± 2). Neuropathology is examined in the offspring on postnatal day 11 and at the termination of the study. The neuropathology analysis includes simple morphometric measurements of brain areas.

Although these studies are designed to specifically assess the effects of developmental exposures on nervous system structure and function, they are limited in the extent to which this complex system can be evaluated as part of routine testing. For example, assessment of social and reproductive behavior and condition (such as anxiety) are not included, different types of learning and memory — such as spatial and sequential learning, reference and working memory, or the effects of recall delay — are not assessed, and long-term effects of developmental exposures (beyond 60 days (d)) are not evaluated. Several efforts are under way to evaluate the utility of such protocols and to improve the methods used in rodent studies so that they are more comparable to those used in humans.

In Vitro Assays

Any developmental toxicity assay that uses a test subject other than a pregnant mammal falls under the general heading of an "in vitro assay." Examples include isolated whole mammalian embryos in culture, nonmammalian embryo culture, and tissue, organ, and cell culture. Several manipulations are possible using in vitro assay systems that are not possible using pregnant mammals, such as the removal of the maternal environment, the removal or transplantation of specific tissues and cells, and the ability to track specific cells and molecules, to genetically alter cells, or to monitor embryo physiology.

There are two potential applications for in vitro assays: screening for developmental toxicity and analyzing mechanisms of normal and abnormal development. In vitro assays to screen chemicals for potential developmental toxicity have been under development for approximately 15 years with the idea that they could be used to assess larger numbers of chemicals than can be evaluated with in vivo developmental toxicity tests in mammals, could reduce the number of experimental animals used in those tests, and could be used to reduce the costs of testing large numbers of chemicals. A number of attempts have been made to validate in vitro assays for screening chemicals, and efforts are under way to validate the rodent embryo culture, micromass, and stem cell assays in a European-sponsored trial (Spielmann et al. 1998), and the frog embryo teratogenicity assay (*Xenopus*) in an interlaboratory comparison (Fort et al. 1998). Validation requires certain considerations in study design, including defined endpoints for toxicity, an understanding of the procedure's ability to respond to chemicals that require metabolic activation, and the accuracy of the test's response to chemicals that cause developmental toxicity or no effect in whole animal studies (Kimmel et al. 1982; Kimmel and Kochhar 1990; Schwetz et al. 1991). Since most in vitro systems involve an interruption in normal metabolism and the biological interrelationships found in the intact system, the range of developmental effects that can be produced and the power of the study to detect an effect are compromised as compared to those obtained using standard study designs in whole animal systems (Kimmel 1990). For these and other reasons, in vitro developmental toxicity assays are unlikely to be used alone to screen chemicals for risk assessment purposes when there is no prior knowledge about the potential for developmental toxicity. In the case of priority-setting in early drug or chemical development, such assays may be useful for eliminating those with toxicity that can be detected in these systems, leading to further development of those with little or no toxicity, with the expectation that standard in vivo assays would be conducted before actual marketing. In vitro screens also may be useful for assessing the developmental toxicity of chemicals or chemical classes for which there is already some information about toxicity from in vivo studies for the purpose of describing the relative toxicity (potency) of members of chemical families. If chemicals are likely to act through a common mechanism, a single in vitro screen that is sensitive to a particular mechanism may predict the relative potencies of a class of chemicals. For example, an in vitro mouse limb bud cell screen has been used successfully to rank the relative teratogenic potential of a large series of synthetic retinoids (Kistler 1987). In addition, in vitro assays may be useful for studying complex mixtures for synergism or antagonism, and for evaluating the cumulative risk of two or more chemicals that have similar mechanisms or effects.

In vitro assays have become widely used for mechanistic studies in developmental toxicology (Harris 1997). An advantage to using in vitro assays for such studies is that they utilize decreasing levels of biological complexity to isolate specific developmental processes. In vitro assays are useful for identification of tissue sites of accumulation, initial biochemical insults, gene expression changes, structure-activity relationships, and disrupted developmental pathways. It is important to link the information developed in these assays to the whole tissue and organism events that are seen as a result of developmental toxicity in order to be most useful for risk assessment purposes. Such information can be employed in developing biologically based dose-response models for developmental toxicity (e.g., Shuey et al. 1994).

Interpretation

Box D-1 lists endpoints that can be used to assess developmental toxicity from standard testing studies. EPA (1991) published guidelines for developmental toxicity risk assessment that provide more detailed discussion of study result interpretation.

Observations on dams during the course of a study include regular examination for signs of toxicity and measurements of body weight. Assessment of food and water intake also can indicate toxicity and is essential to calculate actual test substance intake when the substance is administered in the diet or in drinking-water. When an agent is known to produce pharmacological or toxic effects, including sedation, respiratory depression, or hemolysis, such endpoints also are monitored. Maternal observations assess the relative contribution of maternal toxicity to any embryo-fetal toxicity observed. Maternal body weight before and after removal of the gravid uterus allows the determination of toxicity to the mother exclusive of effects on uterine content.

Examination of the uterus and its contents and of the ovaries of animals that are killed before parturition allows determination of the number of corpora lutea (a measure of the number of eggs released); implantations; live, dead, and resorbing fetuses; fetal weight; and sex. The number of implantation sites equals the number of live fetuses plus the number of dead embryos and dead fetuses. Preimplantation loss can be determined by subtracting the number of implantation sites from the number of corpora lutea. It is possible that the treatment can prevent implantation, and caution should be applied when interpreting the number of implantation sites and preimplantation loss. Dividing the number of resorptions (embryonic deaths) by the total number of implants gives a measure of postimplantation loss, subject to the same caution as above. It should be noted that postimplantation loss is sometimes expressed inclusive of fetal deaths. Uteri that show no signs of implantation at all can be stained with ammonium sulfide to reveal completely resorbed implantation sites (Salewski 1964).

Viable fetuses are examined for external, visceral, and skeletal malformations and variations, and the sex is determined. Individual fetal weight and identification allow external, visceral or skeletal

Box D-1 Developmental Toxicity Endpoints from Standard Testing Protocols

Endpoints typically measured at terminal phase of pregnancy
  Preimplantation loss
  Implantation site
  Corpora lutea
  Resorptions and fetal death
  Live offspring with malformations and variations
  Affected (nonlive and malformed) conceptus
  Fetal weight

Endpoints that can be measured postnatally
  Stillbirth
  Offspring viability (birth, within the first week, weaning, etc.)
  Offspring growth (birth, postnatal)
  Physical landmarks of development (e.g., vaginal opening, palano-
    preputial separation)
  Neurobehavioral development and function (actual enpoints
    measured depend on the function or organ system being studied)
  Reflex development
  Locomotor development
  Motor activity
  Sensory function
  Social-reproductive behavior
  Cognitive function
  Neuropathology and brain weight
  Reproductive system development and function
  Vaginal opening
  Onset of estrus
  Balano-preputial separation
  Ovarian cyclicity
  Quantitation of ovarian primordial follicles
  Sperm measures (e.g., morphology, motility, number)
  Fertility
  Pregnancy outcome
  Other organ system function (e.g., renal, cardiovascular)

Source: Adapted from EPA 1991.

findings to be linked to individual weights. Because there is a correlation between the number of fetuses in a litter and fetal weight, fetal weight can be analyzed with litter size as a covariate.

It is helpful to distinguish between early and late resorptions because dose-related effects can assist in determining the period during development that is sensitive to the test agent. Placental examination and weight might be of value in interpreting results. It is the commonly accepted practice for studies of rats, mice, and hamsters to allocate fetuses alternately for visceral or skeletal examination; this is done when fetal sectioning is used for visceral examination and the fetus cannot be examined skeletally. Where fresh microdissection is used, the fetuses can be examined both viscerally and skeletally, except for the head, which is usually fixed for head sections to examine the brain, eyes, nasopharynx, and other structures or processed with the skeleton to examine skeletal structures. For rabbit or larger fetuses, each fetus usually is examined both for visceral and skeletal effects. Several techniques are used for skeletal examination, including single-staining with alizarin red S, double-staining with alizarin red S and alcian blue to show both ossified bone and cartilage, or X-rays with or without intensification (Inouye 1976; Whitaker and Dix 1979).

Maternal and developmental endpoints are evaluated to interpret developmental toxicity data (EPA 1991). Of particular concern are agents for which there are no signs of toxicity to the maternal animal but that induce toxicity in the developing offspring, or when developmental effects are observed at doses below those causing toxicity in maternal animals. Another common situation is when adverse developmental effects occur only at doses that cause minimal maternal toxicity. In these situations, developmental effects should be attributed to developmental toxicity and should not be considered secondary effects of maternal toxicity. It is possible that the adult and the developing offspring are sensitive to the same dose of an agent. Also, it is important to note that maternal effects might be reversible whereas developmental effects could be permanent. Data on developmental effects can be difficult to assess when they occur at doses that cause severe maternal toxicity.

REPRODUCTIVE TOXICITY

Manifestations

Male Reproductive Toxicity

Expressions of male reproductive toxicity can involve alterations in the male reproductive organs or in related endocrine systems. Such alterations can include changes in sexual behavior (mating behavior, libido, erection, intromission, ejaculation), onset of puberty (delayed physical and behavioral development), fertility (achieving conception within a defined period), pregnancy outcome (production of normal quality and number of offspring), reproductive organ structure and morphology, reproductive endocrine parameters (including peptide and steroid hormone control), or other functions that compromise the integrity of the male reproductive system.

Female Reproductive Toxicity

Female reproductive toxicity includes adverse effects on reproductive organs and related endocrine systems. Endpoints that reflect toxicity include sexual behavior (receptivity to the male at appropriate times in the cycle), age at onset of puberty, fertility (the ability to produce offspring in normal number), gestation length, parturition, lactation, loss of primordial follicles, and age at reproductive senescence.

Types of Studies

Single-Generation Reproduction Study

The single-generation test can provide useful information on basic reproductive function (OECD 1983). It also provides information on the effects of subchronic exposure of peripubertal and adult animals.

In a single-generation reproduction study, males and females can be exposed during the same or separate trials to determine whether one or both sexes are affected. Daily dosing of male laboratory animals should begin when they are 5 to 9 weeks (wk) old and continue for 10 wk (for rats) or 8 wk (for mice) before the mating period. This schedule exposes the animals to an agent for the duration of one complete cycle of spermatogenesis (approximately 70 d in rats and 56 d in mice). Daily dosing of females should begin when they are 5 to 9 wk old and continue for at least 2 wk (OECD 1983) before mating. Females should continue to receive daily doses of the test agent throughout the 3-wk mating period, during pregnancy, and until offspring are weaned. At least three dose groups and one control group are usually included. Either one male to one female or one male to two female matings can be used, resulting in group sizes of at least 20 females and 10 or 20 males. The goal is to produce a minimum of 20 pregnant females per treatment group. Animals that have not mated or that remain infertile should be separated and studied for the cause of their infertility.

Animals are allowed to litter normally and rear their progeny until weaning. Optionally, by the removal of some pups, the litters can be standardized (normally on day 4 postpartum) to include an equal number of pups of each sex (Agnish and Keller 1997). It is considered inappropriate to remove only runts or any other deviant animals. Adjustment of litter size is not possible when there are fewer than eight animals per litter. The major advantage of the standardization of litter size is the diminished variability of pup and litter data, because dams have equal lactational challenges and pups have similar possibilities for growth. The disadvantages of standardization have been documented extensively elsewhere (Palmer 1986; Palmer and Ulbrich 1997) and include the disruption of the normal distribution of litter sizes; standardized litter sizes that are below the natural mean, median, and modal values normally observed for most rat and mouse strains; the elimination of large numbers of offspring that normally would survive; the introduction of human bias in selection; and the raising of mean body weight of pups and the lowering of the challenge to the lactating ability of the dam.

The animals are observed daily throughout testing. Toxic effects, mortality, neurobehavioral changes, altered sexual behavior, and problems in parturition and lactation are recorded. Food consumption and weight of animals are measured at least weekly, and after parturition on the same days litters are weighed. Individual records of each parent test animal and litter are maintained. The time after pairing to achieve a sperm-positive smear (the precoital interval) and duration of pregnancy are recorded, and soon after delivery the number and sex of pups, stillbirths, live births, and the presence of gross anomalies in each litter are recorded. The pups are weighed at a minimum on the morning after birth, on days 4 and 7, and weekly thereafter. Dead pups and the excess pups killed at day 4, if the litter is standardized, should be studied for any defects. All abnormalities in the dams or offspring should be recorded.

At necropsy, the offspring are examined for structural abnormalities, particularly those of reproductive organs, that also can be preserved for histopathological study. At a minimum, all parental animals and offspring that die during the test, those in the highest dose group, and the controls should be examined. Whenever there are gross abnormalities in an organ, the animals also must be examined microscopically.

The data should be treated with appropriate statistical methods. If one male is mated to two females, then nested statistical analysis must be performed based on the number of males used. A well-conducted, single-generation reproductive toxicity study should provide an estimate of a no-observed-adverse-effect level (NOAEL) and an assessment of adverse effects on fertility, parturition, lactation, and postnatal growth. Significant detrimental effects on any endpoint should be considered adverse.

The primary limitation of a single-generation toxicity study is that it provides no information on the breeding capacity of offspring. Other limitations are noted in Table D-1. An EPA workshop (Francis and Kimmel 1988) examined the value of the single-generation reproductive study and concluded that it is "insufficient to identify all potential reproductive toxicants, because it would exclude detection of effects caused by prenatal and postnatal exposures (including the prepubertal period) as well as effects on germ cells that could be transmitted to and expressed in the next generation" (EPA 1996a).

Multigeneration Reproduction Study

Several authorities have published guidelines on multigeneration reproductive assays (OECD 2000b; FDA 2000; EPA 1998b). For a discussion of multigeneration tests, see articles by Lamb (1988, 1989) and Christian (1986).

Multigeneration reproduction studies determine the potential of an agent to produce adverse effects on the male and female reproductive systems, in the embryo and fetus, and in the neonate. The bioassay examines a wide variety of endpoints related to reproduction, including effects on libido; germ cells; gametogenesis; fertilization; implantation; embryonic, fetal, and neonatal growth; development; parturition; lactation; and postweaning growth and maturity. The direct toxic effects of an agent on the pregnant dam can be evaluated. In addition, the recently revised guidelines include measures of estrous cyclicity and ovarian primordial follicle counts in parental and first filial (F_1) females, and sperm parameters (number, motility, and morphology) in parental and F_1 males. Development of the reproductive system and measures of sexual maturation (vaginal opening and preputial separation) are also included. Finally, organ weights of the reproductive organs, target organs, and brain, spleen, and thymus are included. Because of its study design, a multigeneration reproduction study can provide data that cannot be developed from other standard testing protocols. The observed effects can be different from those seen in other (e.g., subchronic) studies.

The parental animals (P generation) are treated with an agent, usually via ingestion, for at least 10 wk before mating. Females continue to be exposed during gestation and lactation. Each dam can produce one to three litters, depending on whether the outcome in the first litter is unequivocal or confirmation is required of findings. This gives some flexibility in the protocol (multiple litters are needed only when initial results are equivocal), and it applies to P and F_1 generations. If the effects in the F_2 (second filial) generation are more marked than in the F_1 generation, additional generations can be examined to clarify potential transgenerational effects.

Dosing of all generations is continuous throughout the study. During lactation, pups receive the test substance through the dam's milk and later from the treated food or drinking-water. If inhalation exposure is used, grooming of the fur can lead to additional exposure to the test material. Coprophagia by the pups is another possible route of exposure. Upon reaching sexual maturity, at least one male and one female from the F_1 generation are selected from each litter for mating with another pup from a different litter but exposed to the same dose.

F_1 generation rats are treated for at least 13 wk and F_1 mice are treated for at least 11 wk before mating.

The study report must include the following data:

- Species and strain.
- Toxic response data by sex and dose, including indices of mating, fertility, gestation, birth, viability, and lactation; offspring sex ratio; time-to-mating (including the number of days until mating and the number of estrous cycles until mating); duration of gestation.
- Day of death if that occurs during the study.
- Toxic or other effects on reproduction and pre- or postnatal growth of the offspring.
- Developmental data, such as anogenital distance (triggered in F_2 pups if positive findings are noted in the F_1 animals), age of vaginal opening, and preputial separation.
- Number of P and F_1 females with normal cycles and cycle length.
- Day of recording an abnormal effect and its subsequent course.
- Body weight data by sex for each generation.
- Dietary intake and food efficiency (body weight gain per food consumed), and test substance consumption for P and F_1 animals, except for the period of cohabitation.
- Sperm evaluation on data including total cauda epididymal sperm counts, percentage of progressively motile sperm, percentage of morphologically normal sperm, and percentage of sperm with each identified abnormality.
- Stage of estrous cycle at the time of death for P and F_1 females.
- Necropsy findings.
- Implantation data and postimplantation loss calculations for P and F_1 females.
- Absolute and relative organ weight data.
- Detailed descriptions of all histopathological findings.
- Adequate statistical analyses.
- A copy of the study protocol.

The multigeneration study is probably the most complex type undertaken for regulatory purposes and provides information on toxicity that follows treatment throughout the entire reproductive cycle, except that it does not evaluate reproductive senescence other than the evaluation of primordial follicles in females. Other limitations are noted in Table D-1. In general, significant detrimental effects on any endpoints or on indices derived from the data should be considered adverse. EPA (1996a) provides a detailed discussion on adverse effects.

Reproductive Assessment by Continuous Breeding Study

The U.S. National Toxicology Program (NTP) has developed a test protocol for evaluating toxicity through a reproductive assessment by continuous breeding (RACB) study design (Lamb 1985; Gulati et al. 1991). The protocol was originally developed for mice as a faster and more cost-effective alternative to the conventional regulatory reproductive toxicity studies, but it also has been used successfully with rats. After a 1-wk pretreatment period, males and females are housed as breeding pairs in individual cages and allowed to mate continuously for 14-wk. Exposure to the test substance (usually in feed or drinking-water) is continued throughout the study, and the offspring are removed from the cage immediately after parturition. After the cohabitation breeding period, the pair is separated and the last litter is raised to weaning. Pups from these litters can then be selected, and treatment is continued. The pups are used in a mating trial to evaluate effects in the second generation in a manner comparable to that described previously for the multigeneration design.

The same endpoints (fertility, pups per litter, pup weight, sex, survival) are studied in the RACB protocol as in the standard multigeneration protocol. It is the time between litters and the progressive effects on fertility and reproduction that are specific to the RACB study design. The difference is that the RACB study design allows more than one litter to be examined per generation and can give an indication of subfertility and infertility. Adverse effects that might not be noted in the first mating may become evident later due to longer exposure time; such findings would not normally be detected in the conventional studies.

The RACB does not give information on specific male and female reproductive effects unless cross-breeding of control and treated males and females is done following the 14-week mating trial. It also does not provide information on effects in the second generation unless F_1 pups are raised to breeding age and mated to produce a second generation as in the multigeneration study design. Other limitations are noted in Table D-1.

Serial Mating Study (Dominant Lethal Study)

If a single-mating trial results in an adverse effect attributable to the male, it is difficult to determine the developmental stage in which the disruption occurs. It is well known that different stages of spermatogenesis are variably sensitive to toxic effects and that each toxic substance can affect different sperm cell populations (Parvinen 1982). Spermatogonia, for example, are sensitive to cyclophosphamide in experiments conducted in mice (Toppari et al. 1990), whereas spermatocytes are disrupted by ethylene glycol monomethyl ether in experiments conducted in rats (Chapin et al. 1985). The action of a compound that primarily affects the somatic Sertoli cells of the testis, for example, m-dinitrobenzene (Foster 1989), will produce an extensive period of infertility because of adverse effects on the function of these cells at various stages of germ cell differentiation.

Serial mating makes it possible to assess the sensitive stages of spermatogenesis and susceptible cell types. This information can be obtained from a specific serial-mating trial or from a similar protocol used for dominant lethal testing (OECD 1984; EPA 1998d). Adult males (usually rats) are exposed before mating, typically for 1-5 d, with 20 males per dose group, where after they are mated to one to three females weekly for the next 8-10 wk. Adverse effects on male reproduction are manifested as decreased numbers of implantation sites in uteri (indicative of failure of fertilization or preimplantation loss) and increased early fetal mortality (indicative of postimplantation loss or dominant lethality). To examine the uterine contents, dams are killed before parturition (e.g., on days 13-18 of pregnancy).

Any adverse effects can then be attributed to specific cell populations by back-calculation on the basis of the well-known kinetics of spermatogenesis (Chapin et al. 1985). The test was originally designed for detection of germ cell mutagenicity, and it requires a large number of female animals (e.g., an 8-wk trial would use 160-480 females), which is a disadvantage.

The limitations of serial-mating trials are similar to those shown in Table D-1 for other reproduction studies, except for the identification of stage of spermatogenesis affected. Additional endpoints of male reproductive toxicity and effects other than death of the offspring are not evaluated unless included in the protocol.

Total Reproductive Capacity

The total reproductive capacity study, a variant of the continuous-breeding study, is designed to assess ovarian toxicity. Female fetuses are particularly susceptible to agents that can adversely affect germ cells because development of the oocyte occurs prenatally; no new germ cells develop after birth. Female animals are exposed to a test substance for a short period in utero (i.e., days 9-16 of gestation) (McLachlan et al. 1981) or postnatally (Generoso et al. 1971) and allowed to mate with a single male as long as the females remain fertile. The numbers of litters and offspring are compared with those of control animals to estimate the loss of oocytes resulting from the exposure.

Total reproductive capacity studies have been designed with the specific purpose of evaluating female reproductive capacity and are not tests of general reproduction function.

Interpretation

Well-conducted multigeneration and continuous-breeding studies can provide data that demonstrate changes in the key parameters of male and female fertility and reproduction. Statistically significant, dose-related changes in the indices listed in Table D-2 provide sufficient evidence of reproductive toxicity but by themselves do not identify the affected sex. Because most multigeneration or continuous-breeding studies place test males with females treated at the same dose, they cannot identify which sex is affected. Although such studies are the most typical way to evaluate the reproductive toxicity of an agent, most provide insufficient evidence of whether the agent causes male or female reproductive toxicity in animals. There is, therefore, a need for additional data, which, in fact, can come from the same study. For example, evidence of gonadal toxicity measured by testicular

TABLE D-2 Indices of Fertility and Reproductive Function

Index	Calculation
Female mating index	$\dfrac{\text{No. estrus cycles with copulation}}{\text{No. cycles required for conception}} \times 100$
Female fertility index	$\dfrac{\text{No. females presumed pregnant}}{\text{No. females cohabited}} \times 100$ or $\dfrac{\text{No. females inseminated}}{\text{No. animals paired}} \times 100$
Female fecundity index (also, called the conception or pregnancy index)	$\dfrac{\text{No. confirmed pregnant}}{\text{No. with copulatory plug or sperm}} \times 100$
Male mating index	$\dfrac{\text{No. males with pregnant females}}{\text{No. males}} \times 100$
Parturition index	$\dfrac{\text{No. parturitions}}{\text{No. females confirmed pregnant}} \times 100$
Gestation index	$\dfrac{\text{No. females with pups born alive}}{\text{No. confirmed pregnant}} \times 100$
Live litter size	$\dfrac{\text{No. live offspring}}{\text{No. females with copulatory plug or sperm}} \times 100$
Live birth index	$\dfrac{\text{Mean pups per litter born alive}}{\text{Mean pups per litter delivered}}$
Viability index	$\dfrac{\text{Mean pups per litter alive day 4}}{\text{Mean pups per litter born alive}}$
Lactation index	$\dfrac{\text{Mean pups per litter alive day 21}}{\text{Mean pups per litter alive day 4}}$
Weaning index (if litter size artificially reduced)	$\dfrac{\text{Mean pups per litter alive day 21}}{\text{Mean pups per litter kept at day 4}}$
Preweaning index	$\dfrac{\text{Mean pups born per litter minus mean pups weaned per litter}}{\text{Mean pups born per litter}}$

weight or altered morphology can provide sufficient evidence that an agent is a male reproductive toxicant or add weight to evidence that it is not a male reproductive toxicant. Likewise for females, evidence of ovarian toxicity measured by weight changes and altered morphology can provide sufficient evidence for female reproductive toxicity. Another way to provide sufficient evidence of male reproductive toxicity would be to mate the treated animal of one sex to the untreated animal of the other sex.

Male Indices

Organ Weight

A statistically significant, dose-related decrease in absolute or relative testicular weight is generally sufficient evidence that an agent can cause reproductive toxicity in animals. Most agents that cause testicular toxicity also cause decreases in testicular weight, but if they cause edema, the testicular weight increases. Decreases in testicular weight can be considered sufficient evidence of toxicity by themselves, but increases must be explained by other endpoints, such as morphology. Any changes also must be considered in light of the systemic toxicity elicited by the test chemical. Severe systemic toxicity brings into question not only the organ weight data, but also the relevance of any other reproductive effects.

Weight changes in male accessory sex organs can indicate significant functional effects. Both the seminal vesicles and the prostate, for example, contain a large proportion of luminal fluid that can decrease rapidly when androgenic hormone concentrations decline. Epididymal weight is largely affected by the number of sperm present in the epididymis. Statistically significant, dose-related decreases in the weight of the epididymis would be sufficient evidence of male effects. Decreases in the weight of the seminal vesicles or ventral prostate can be sufficient evidence of male reproductive toxicity, but are more useful if supplemented by data on endocrine effects. Changes in pituitary weight alone would typically be insufficient evidence of male reproductive toxicity, because pituitary weight is an inaccurate indicator of changes in pituitary function, which are best measured by other parameters, such as hormone concentrations. Furthermore, only a small portion of the gland is involved with reproductive function.

Organ Morphology

Changes in testicular morphology are best observed when the tissues are preserved by optimal methods. The best evaluations can be done on testes fixed by perfusion and embedded in a plastic, such as glycol methacrylate. More conventional, but still quite acceptable, morphologic investigations can be performed on testes fixed by immersion in Bouin's fixative, embedded in paraffin, and stained with PAS. Formalin fixation and paraffin embedding of testes is an inferior and generally inadequate method for the study of testicular pathology because it will reveal only the most severe effects. In formalin-fixed and paraffin-embedded tissues, only the most severe changes in the seminiferous epithelium of the testis could be considered sufficient evidence of male effects. The sensitivity of these evaluations can be substantially improved by more careful fixation, embedding, and observation techniques. Low-quality morphological techniques, such as formalin fixation and paraffin embedding, are never sufficient to show that an exposure did not produce testicular toxicity.

Morphological changes in accessory sex organs are less common, but clear treatment-related effects also can provide sufficient evidence of male effects.

Sexual Behavior

Fertility studies do not incorporate measures of sexual behavior, but they indirectly measure endpoints that can be altered by effects on sexual behavior. These measurements include collecting vaginal smears to check for the presence of sperm or checking vaginal plugs as evidence of mating. An azospermic male, however, might have normal sexual behavior but will not have a "sperm-positive" mating. Thus, even though a decrease in sperm-positive matings can be sufficient evidence of reproductive toxicity, it would not be sufficient evidence of abnormal sexual behavior. If a study does measure sexual behavior, mounting frequency, intromission, ejaculation number, and latency can be measured. More detailed studies of sexual behavior (Zenick and Clegg 1989) would be helpful, but are rarely done.

Sperm Evaluation

In mice and rats, sperm motility and count are relatively sensitive and reliable indicators of male reproductive toxicity (Morrissey et al. 1988a,b). Statistically significant, dose-related decreases in these parameters would constitute sufficient evidence of male reproductive toxicity, even if fertility is not adversely affected. Sperm morphology changes, if statistically significant and dose-related, would be sufficient evidence of reproductive toxicity. Experience has shown, however, that sperm morphology changes in rodents are fairly insensitive indicators of reproductive toxicity (Morrissey et al. 1988a,b) even though they can be good indicators of reproductive dysfunction in humans.

Sperm evaluations in rats and mice are nearly always limited to the terminal sacrifice of the test animals because it is extremely difficult to collect semen samples from such small animals. Because investigators can collect whole semen samples from rabbits and domestic animals, however, it is possible to assess and follow progressive changes in semen in these animals over time. The potential advantages to conducting sperm assessments in rabbits include the ability to assess the same parameters (morphology, motility, sperm count) at successive points. Studies have shown that large decreases in semen parameters must occur before there are noticeable changes in fertility. Statistically significant, dose-related decreases in semen quality, however, could constitute sufficient evidence that an exposure causes reproductive effects in the test species.

Endocrine Evaluations

If adequately designed studies detect changes in concentrations of gonadal steroid or gonadotropic pituitary hormones, these endocrine parameters do provide sufficient evidence of reproductive toxicity. Typically, adequate studies that show toxicity will have multiple samples obtained in a well-defined context that includes sex, age, reproductive state, day of cycle, and so on. Endocrine changes that indicate toxicity will include both multiple values outside the normal physiological ranges and physiologically plausible changes in direction in hormone concentrations.

Biochemical Markers of Reproductive Exposure and Effect

Various markers of exposure and effect have been investigated in male reproductive toxicology, including prostatein, androgens, and prolactin (NRC1989). Sertoli cell enzymes or biochemical secretory products, measured in vitro and in vivo as markers of cell function, are other examples of useful endpoints for studying target organ or cell responses. Currently, however, they cannot be considered evidence of male reproductive toxicity.

In Vitro Methods

There are methods for culturing various cells from the male reproductive system, such as pituitary cells, Sertoli cells, and germ cell-Sertoli cell cocultures. Although these investigations help elucidate mechanisms of action, they cannot by themselves generate sufficient evidence of reproductive toxicity.

Female Indices

Several endpoints listed in Table D-2 can provide evidence for female reproductive toxicity. For example, when a continuous-breeding study shows an adverse effect, it is desirable that the study also mate each member of a breeding pair to an untreated control to identify which member is affected by the agent. If a study has not taken this step, it cannot be said with certainty that the observed effect is the result of female reproductive toxicity; it can be equally likely that a male effect or a couple effect is involved.

Because most standard animal reproduction studies do not observe mating, they do not contain evaluations of an agent's effect on sexual behavior. If a study does report observations of mating, the failure of female rodents to assume a lordotic position and to accept mounting is evidence of abnormal sexual behavior. Additional signs include running from or fighting with the male (Uphouse and Williams 1989; Uphouse 1985).

Cytology Abnormalities

Abnormal findings for estrous animals include persistent estrus, prolonged diestrus, or anestrus (May and Finch 1988). To characterize the estrous cycle in appropriate experimental animals, studies can use vaginal cytology or other cyclic signs in animals that menstruate, including humans. These parameters can give information on whether cycling has discontinued or whether segments of the cycle are altered in length. Because estrous cycle length has a normal variation, it is also possible to evaluate changes in the distribution of cycle lengths. The interpretation of these data is, however, open to question. Vaginal cytology data can also be incorporated into such protocols as the continuous-breeding test, the subchronic study, and the two-generation reproduction study (Morrissey et al. 1988a,b; EPA 1998b; OECD 2000b; FDA 2000). Alterations in the distribution of estrous or menstrual cycle length alone have not been shown to be reliable predictors of reproductive toxicity. By themselves, these alterations would be insufficient to identify an agent as a reproductive toxicant.

Weight and Morphology Changes

A statistically significant decrement in ovarian or uterine weight in a study properly controlled for cyclic variation is worthy of consideration and should signal the need for additional studies. Similarly, an increase in uterine weight in an acyclic or castrate animal, or in a study that controls for cyclic variation, should raise concern about possible estrogenicity of the test agent and should suggest that additional studies are needed. Neither of these parameters, as an isolated endpoint, is sufficient to characterize an agent as a reproductive toxicant. Evaluation of the ovary often includes counts of follicles or subpopulations of follicles (Pederson and Peters 1968; Heindel 1999). A decrease in the number of ovarian follicles or a change in follicle subtype, however, is evidence of reproductive toxicity.

Biochemical Changes

Secretion products of the uterus can be obtained with uterine lavage (Teng et al. 1986). Changes in uterine secretions could be useful for characterizing alterations associated with treatment because these changes can be cycle dependent, however, they can be difficult to interpret. To date, the characterization of normal changes in uterine secretory products is incomplete. Such changes alone, however, are insufficient to characterize an agent as a reproductive toxicant.

Alterations in Age at Puberty or Reproductive Senescence

In animals with estrous cycles, the onset of puberty is marked by vaginal opening. Reproductive senescence may manifest as persistent vaginal estrus followed by anestrus. A change in the age at puberty or reproductive senescence is sufficient to characterize reproductive toxicity, although it is desirable to have supporting data that explain the mechanism of toxicity.

Endocrine Parameters

In estrous and menstrual animals, the reproductive cycle is characterized by the production of sex steroids from the ovary in response to pituitary gonadotropins, which are under hypothalamic control. It is possible to measure the relevant hormones, but evaluators must keep in mind that the hormones are produced in a pulsatile fashion, with cyclic variation in the amplitude and frequency of the pulses. For this reason, single static measures are unlikely to be informative unless a result is well outside the normal ranges (e.g., castrate concentrations of gonadotropins). Other strategies for evaluating endocrine parameters include serial measurements of hormones in blood at short intervals, and response of an endocrine measure to a stimulus. In the serial measurement strategy, frequent sampling permits the construction of a profile of the hormone change over time, which can disclose the pulse pattern. This method is difficult in animals with small blood volumes where frequent sampling may produce its own effects.

The second method, response of an endocrine measure to a stimulus, involves sampling an animal at a fixed time after administration of a releasing factor. One can, for example, measure luteinizing hormone after injecting gonadotropin-releasing hormone or measure progesterone after injecting chorionic gonadotropin (Hughes 1988). The disadvantage of this method is the possibility that the injection of the releasing agent will cause an atypical physiological situation, so that one cannot extrapolate the effect it "unmasks" to unmanipulated animals.

If changes in concentrations of gonadal steroid or gonadotropic pituitary hormones are detected in adequately designed studies, these endocrine parameters do provide sufficient evidence of reproductive

toxicity. Results that show multiple values outside the normal physiological ranges, changes in hormone concentrations in physiologically
plausible directions, or failure of key hormonal events (such as luteinizing hormone surge, preovulatory estradiol rise, maintenance of luteal
phase progesterone production) provide sufficient evidence of reproductive toxicity.

In Vitro and Perfusion Systems

Tissue culture methods have been used to study ovary slices in
vitro, and cell culture methods have been used for studying granulosa
cells and myometrial cells. In culturing ovary slices or granulosa cells,
investigators often use the release of sex steroids into the medium as
an outcome parameter. Under some conditions, granulosa cells will
luteinize, producing a range of steroid and nonsteroid products; of
these, progesterone is measured most commonly. Some studies, however, have measured other products, including nonsteroidal substances
(Haney et al. 1984; Teaff et al. 1990). Some cell culture studies have
made use of the contractile properties of myometrial cells for evaluating the potential of agents to alter uterine activity. In all of these test
systems, the artificial nature of the in vitro setting can limit the predictive value of the results.

Ovaries perfused in vitro are useful systems for studying the
mechanical aspects of ovulation. The preparations allow observations
on the effects of agents in preventing rupture of the follicle and expul-
sion of the oocyte. The perfusion system is artificial, however, and the
relocation of the ovary from peritoneal cavity to the perfusion chamber
can alter the mechanical features of the system. For this reason, data
from perfusion studies are not, in themselves, sufficient for drawing
conclusions about an agent's reproductive toxicity.

Any change observed in an in vitro or organ perfusion system
should be considered supplemental. Isolated findings of studies that
use these systems are insufficient to characterize an agent as causing
reproductive toxicity.

Breast Milk

Changes in breast histopathology or in breast milk amount or composition should signal the need for additional studies, and in particular, the need for studies that evaluate the effect of such changes on the nourishment and health of the offspring. The mere presence of xenobiotics in milk is not, by itself, evidence of toxicity; however, if a test agent is concentrated in milk, this should prompt recognition of the need for studies on the nursling. Conversely, if an agent is not transferred into the milk in rodent studies, but it is clear that exposure to critical organ systems continues in utero at the same developmental stages in humans, it may be appropriate to conduct direct dosing studies in rodents to determine any potential effects on the structural and functional development of these systems.